WHY BOTHER WITH MATHS

WHY BOTHER WITH MATHS

Paul F. Easthope PhD

Copyright © 2024 by Paul F. Easthope. All rights reserved.
ISBN 979-8-3408-0584-3

For Jane and Iona.

Contents

1 PREFACE 1
2 PRELIMINARIES 7
3 TRIGONOMETRY 11
4 ALGEBRAIC EXPRESSIONS 19
5 SIZE OF A KILLER ASTEROID 27
6 PROBABILITY 29
7 CALCULUS 43
8 BICYCLE WHEEL REFLECTORS 55
9 LADDER IN A MINE SHAFT 59
10 LENGTH OF A SPIRAL 65
11 NEWTONIAN DYNAMICS 69
12 MOMENTS AND BALANCE 81
13 WORKING OUT AREAS 93
14 VECTORS (AND RELATIVES) 99
15 CIRCUITS AND RESONANCE 109
16 FREQUENCY ANALYSIS 117
17 COMPLEX NUMBERS 121
18 QUATERNIONS 129
19 POPULATION DYNAMICS 135

20	INTEGRAL TRANSFORMS	141
21	KALMAN FILTERS	149
22	RANDOM WALK	163
23	CALCULUS OF VARIATIONS	167
24	ANALYTICAL DYNAMICS	177
25	ASSORTED ENGINEERING APPLICATIONS	181
26	CHAOTIC (BUT NOT RANDOM)	199
27	TSUNAMI WAVES	211
28	SPECIAL RELATIVITY	215
29	GENERAL RELATIVITY	229
30	QUANTUM MECHANICS	241
31	OPTIMAL CONTROL	249
32	Acknowledgements	255
	References	256
A	Pythagoras' Theorem	265
B	Medical Test Probabilities	267
C	Expression for Pendulum Period	269
D	Solution for Heavy Chain	273
E	Solution of Quadratic Equations	275
F	Probability Density Mappings	277
G	Green's Theorem in the Plane	279
H	Inverse Laplace Transform	281
I	Projection Slice Theorem	283
J	Random Walk Probabilities	287
K	Bending Beams	293
L	Derivation of Shallow-Water Equations	299

M Parallel Displacement 303

N Greek Letters in Maths 311

Chapter 1

PREFACE

> Somehow, it's okay for people to chuckle about not being good at math. Yet, if I said "I never learned to read," they'd say I was an illiterate dolt.
>
> Neil deGrasse Tyson

> We will always have STEM with us. Some things will drop out of the public eye and go away, but there will always be science, engineering, and technology. And there will always, always be mathematics.
>
> Katherine Johnson

> Mathematics is not only real, but it is the only reality. That is that entire universe is made of matter, obviously. And matter is made of particles. It's made of electrons and neutrons and protons. So the entire universe is made out of particles. Now what are the particles made out of? They're not made out of anything. The only thing you can say about the reality of an electron is to cite its mathematical properties. So there's a sense in which matter has completely dissolved and what is left is just a mathematical structure.
>
> Martin Gardner

THE genesis of this book began with a recent chance remark in conversation, in which the speaker was mentioning his experiences learning mathematics in school: "What is the point of doing all these examples? What use is learning algebra (or calculus, or geometry, or ...)?". This reminded me of numerous instances when relations, or parents of friends, asked what I was studying in school or university. On being told, 'mathematics', the response invariably was along the lines of "Oh, I was never any good at maths in school" (a depressingly common statement [1]), usually uttered in a suitably dismissive tone, which then generated sage nods of agreement from among the assembled company, accompanied by pitying looks in my direction. The implication was, "I've done perfectly well all my life without mathematics and don't need to bother myself with it". This is an attitude that is probably more typical of Western than Asian cultures, since the situation is very different in China, Japan or Singapore, for instance.

It is, in many ways, a rather depressing state of affairs, since mathematics in its various manifestations is virtually ubiquitous in modern life, albeit

1

largely hidden to most people. Virtually every aspect of modern technology — especially when involving microprocessing — requires mathematics. One can mention all the absolutely essential uses of mathematics in physics and engineering, some of which are discussed in sections below. The medieval cathedral builders relied on experience plus a fair amount of trial and error, which is hardly a satisfactory approach for modern engineering; can one imagine building a passenger plane without a detailed understanding of the strength of materials plus the stresses and strains involved? (a question that becomes thoroughly personal when experiencing heavy turbulence).

It is also quite impossible to imagine doing physics without the relevant equations, and this is not just a matter of gaining some understanding of what has already been observed — in effect, just making sense of experiments. There are many examples in physics where a new-found equation has made predictions regarding what *should* be observed, in addition to what has already been seen. Well-known examples here are from general relativity, in which the Schwarzschild metric implied the existence of black holes, which are now accepted as real entities (if not very near at hand, thankfully); or from the Dirac equation in quantum mechanics, predicting the existence of positrons, later confirmed experimentally. See Chapters 29 and 30 for a bit more on these aspects.

Mathematics has been referred to as 'the language of physics', although it is is quite distinct from human languages, in that the corpus, once proved, agreed and established, remains stable thereafter. In contrast, human languages are similar to life itself, being subject to mutations over time (and generally tending towards mutual unintelligibility). This raises the centuries-old question as to whether maths is invented (by humans) or discovered. I do not intend to go any further into this matter, although (to me) the very stability of mathematical results and its outstanding ability to describe physical reality argues for discovery rather than invention; see [2], [3] for more in-depth discussions.

Regarding the mathematical *notation*, however, this is the nearest we have to any sort of international language. Academic or technical documents in Chinese or Russian, for example, implement the same math symbology as in English. The accompanying explanatory text may be in Chinese or Russian, but any mathematician can read and understand the equations. Whether the same goes for extra-terrestrial intelligences is rather more debatable, though.

Then we come to the more immediate uses of mathematics in one's personal life: one need look no further than the mobile phone. Using GPS[1] to determine its position relative to the earth's surface requires complex geometry, in addition to vital timing corrections from relativity; transmitting a conversation needs data compression via Fourier series; data encryption typically uses public-key cryptography which relies on large prime numbers; and so on. It is certain that the whole edifice of digital processing would collapse without Boolean[2] algebra: this is an algebra of ones and zeros, peculiarly suited to digital processing, but predating microprocessors by at least a century. Or look at the aspect of medical imaging which relies on the Radon Transform (Section 20.2), defined well before computers made its application practical in this context. And there

---

[1] Global Positioning System.
[2] George Boole, 1815 to 1864.

are the numerous benefits that math understanding can provide at a personal level, some instances of which may be found in later sections.

A common modern misperception finds voice in the statement "but isn't everything done by computers anyway?", with the expectation that computers have somehow made mathematics redundant. Well, no, thankfully. Computers certainly do the grunt work in CAT (Computerised Axial Tomography) scanners, but at the base of the process is an equation which then needs a mathematically-literate programmer to define the computer tasks. A recent 'computers do everything' misconception is associated with 'big data': accumulate vast amounts of raw data, let the computer loose on it and all sorts of answers and patterns will pop out the other end. Indeed, patterns will be found, but are they real or could they represent random assemblages? An application of probability theory will probably tell you, but big data won't know (Section 6.1 goes a bit further into this issue). The term 'big data' is also currently used in the context of algorithms for selecting among job candidates, allocating loan monies, assessing prisoners for parole, to mention only a few applications; the frequently malign consequences of which are well-documented in [4].

In any case, as pointed out in [5], when journalists and others refer to 'computers' or 'algorithms' being used to solve problems or provide answers, what they invariably mean is mathematics — but dare not say so (for unspecified reasons). On the other hand, computers have certainly proved to be a definite asset in solving complex equations that are not amenable to an analytical approach, and in running randomised trials (as in Section 6.4) — but computers augment maths rather than replacing it.

So to me, at least, the relevance of mathematics is without argument. But this does not answer the obvious question: why should I bother with it, if there are so many other clever people willing to involve themselves instead? One pertinent answer to this comes from medical tests: if you want to understand the probabilities involved — from a personal standpoint — you need to engage with Bayes' theorem (Section 6.2). It is unlikely that anyone will do the work for you, and you may well end up being disadvantaged as a consequence. It is also sobering to think of the many miscarriages of justice that have occurred simply because jurors (and legal minds) were unfamiliar with even basic probability theory (see Section 6.3). However, to go back to the above question, turn it on its head: why should one *not* bother? Sure, the processing in a mobile phone goes on 'under the hood', so to speak, requiring no understanding from the user, but isn't it better to have some appreciation of what is involved, if only to regain a measure of control (and to satisfy one's curiosity)? Otherwise, one may as well be dealing with magic. Besides, maths provides a whole new world to explore, with an astonishing variety of landscapes and terrains, and an uncanny ability to mirror the familiar (or not so familiar) physical world to considerable accuracy using only a few abstract symbols.

Mathematics is often considered 'hard', although — in my view, at least — rather less difficult than some crosswords, especially the 'cryptic' kind. Maths is logical (unlike the crosswords) and, once some understanding is gained, it is much easier to follow the arguments and chains of reasoning. There is also immense satisfaction obtained once a strenuous piece of mathematical reasoning has been completed successfully, perhaps resulting in a single compact equation and vindicating a numerical simulation or matching experimental evidence or suggesting something new to look for. (On the other hand, belatedly finding an algebraic error on line two, thus invalidating several subsequent pages of calculation, is distinctly annoying. This is admittedly one area where computers can help: packages that are capable of symbolic manipulation are much less likely to make such errors than I tend to).

We can legitimately ask, are the recent Asian economic success stories due at least in part to their respect and enthusiasm for technical subjects, of which mathematics is a necessary part? Certainly, these countries generate far more engineers per year than is the case for Europe or the United States, even leaving out population disparities. I do not intend to get on a soapbox and belabour this point, since UK politicians (at least) say they are well aware of the issue (and have been making intermittent noises about it for several generations).

It is probably easier to say what this book is not, rather than what it is. It is not a mathematics primer, as these tend to be directed at specific age groups or curriculum levels. It is not a text book, since many such excellent references already exist. It is not a collection of puzzles, although some of the examples used herein do require a modicum of concentration to understand. Rather, I have elected to include an assortment of examples of mathematics that have been used, at various times through history, to provide answers to practical problems faced by humanity and in attempts to understand the world around them. These examples form a purely personal choice, and a vague apology is made in advance to any fellow mathematicians who may lament the absence of their own topic of expertise. It should be said that the majority of the examples can be found in other text books, although not — to my knowledge — all in one place. Should the reader be interested, the list of references at the end provides an assortment of more standard textbooks, academic papers and links to web sites. In the interests of textual continuity and to avoid unduly distracting the reader, these references are cited in the text in a deliberately minimalist fashion (as in [6]).

As to the expected capability of the reader, no particular assumptions have been made, although some basic understanding of algebra might be beneficial; but not essential, since a brief overview is provided in Chapter 2. The intent is to provide something readable and (hopefully) enjoyable for virtually any age group from early 'teens onwards, so explanatory material is generally more extensive than would normally be the case (as, for example, in a mathematics textbook, where assumptions need to be made to avoid undue repetition). For the same reason, I have attempted to put the more fundamental material toward the beginning of the book, so that later sections can draw on earlier ones.

Don't let anyone tell you that you can't 'do math' — especially if you are a girl. *Anyone* can do math; all that is required is an open mind and some persistence.

No apologies are made for the inclusion of equations; these are the life-blood of maths and to leave them out would render this book largely pointless[3]. Rather

---

[3]It is sometimes said that even the ghost of an equation appearing on a page will scupper

than dump foundational equations on the reader, and expect him or her to accept them on trust, I have tended here to derive such results from scratch (at least, most of the time). This approach serves to illustrate that very simple principles — almost 'common sense' — combined with logical arguments can result in profound results with wide applicability, and in some cases with potentially startling conclusions. Indeed, looking back through the text, it is now obvious to me how frequently Newton's kinematic principles found application to just about anything that moves (as well as a lot of things that don't). I should also have anticipated how foundational and ubiquitous is Pythagoras' theorem, but that will become apparent.

The manuscript was prepared initially using MacTex on an Apple iMac, since TeX and LaTeX are by far the best means of typesetting equations; I am most definitely not in favour of the 'hunt and punt' approach used by some other software products, which becomes excessively tedious after a more than a few minutes (it would certainly have driven me mad in very short order). The graphics have mostly been prepared using `gnuplot` (familiar to many Linux users but equally at home on a Mac), while numerical examples and solutions have, where needed, been created using C code.

I can say that this book has been most enjoyable to write, and certainly the opposite of drudgery. In the process, I've even learnt new stuff, which is supposed to be good for the brain (especially at my age, and which may go some way toward counteracting the adverse effects of the biscuits and cookies consumed in the said writing). There is also considerable satisfaction obtained simply by typesetting equations, which has something to do with the sheer elegance of the composite symbols, quite apart from knowing what the equations mean. Enthusiasts for different fonts will understand this, as will the proud wearers of T-shirts with the equations of general relativity or quantum mechanics across the front.

Mathematics is intellectually satisfying, even leaving aside the enormous reach of the practical side, but you may judge both aspects for yourselves.

The Author, Autumn 2024

---

book sales, but I'm choosing to ignore that forecast.

Chapter 2

PRELIMINARIES

SOME notational preliminaries: the reader may already be familiar with the usual symbols for addition, subtraction, multiplication and division, plus the equals sign, namely: $+$, $-$, $\times$, $\div$ and $=$. Thus, an expression such as $25 \times (87.2 - 34.9) \div 2.6$ is readily understood, noting that the brackets require that particular grouping to be carried out first — it forms a unit.

At its most basic, algebra substitutes letters, such as a, b, c and d, for the numbers, resulting in $a \times (b - c) \div d$. This expression can then represent many different combinations of numeric values. In the same way, each such letter may itself stand in for a group of other quantities, and so enable the construction of more complex equations.

In contrast to the situation in many computer programming languages, the different types of bracket or parenthesis used in expressions and equations — such as ( ), { }, *etc* — generally have no particular significance apart from being employed in (mostly) matching pairs[1]. Such brackets are usually combined in different ways and in different sizes to aid readability and highlight patterns, if any.

Compactness and the need for easy readability have also led over the years to the replacement of an expression such as

$$a \times (b - c) \div d$$

with

$$a(b-c)/d \quad \text{or the equivalent} \quad \frac{a(b-c)}{d},$$

(although the $\times$ symbol may still be needed when concatenating several numbers, such as 24.7×8.6).

Thus, *cat* does not refer to a small furry animal with whiskers, but to a more compact and readable form of $c \times a \times t$ (although such combinations of symbols would generally be avoided in case of confusion). An exception to this

---

[1] An exception is the Dirac notation used in quantum mechanics, in which quantities such as $|\psi\rangle$ have specific meanings.

statement occurs in calculus, where quantities such as dx stand for 'a small bit of x' and have a unitary meaning (see Section 7 for more explanation).

In most cases, the order in which additions, multiplications and divisions are carried out is immaterial, so that $a + b = b + a$ and $ab = ba$. An exception concerns matrix multiplication, which will be met in Chapter 14 (and the non-commutative algebras have interesting properties of their own, which are not really addressed in this book but become important in quantum mechanics, as one application).

There is also the distributive law, which says that $a(b+c) = ab + ac$, which is often used to collect terms together. Some examples of this will appear later.

In an equation involving an equality, the equation as a whole is unchanged by adding or subtracting, multiplying or dividing by any quantity — provided these operations are applied to both sides. For example, with an equation such as $y - a = 2x$, adding a to both sides transforms it to $y = 2x + a$. In a similar manner, $y/a = 3x$ has the same meaning as $y = 3ax$. Chapter 3 below provides some more comprehensive examples. On the other hand, *inequalities*, such as $y < ax + b$, require rather more care (we'll come to that a bit later on as well).

The use of the Greek alphabet, such as the letter α ('alpha'), occurred when the Roman alphabet proved inadequate to cope with all of the letters that mathematicians needed to use. Some Greek letters, such as π, have acquired a more or less standard numerical value (3.141592654... — an irrational number with, so far as anyone knows, a never-ending sequence of digits with no discernable pattern). Appendix N provides a list of the Greek letters used in mathematics, while more recently, even Hebrew letters have been pressed into service, with ℵ, 'aleph', being perhaps the best known.

There is also a whole menagerie of symbols (other than the familiar algebraic ones above), some of which will appear later in their appropriate contexts.

Finally, a word on coordinates: to locate a point on a line, only a single distance from some reference point is required (think distance along a ruler). For points on a plane, two numbers are needed, as illustrated in Figure 2.1.

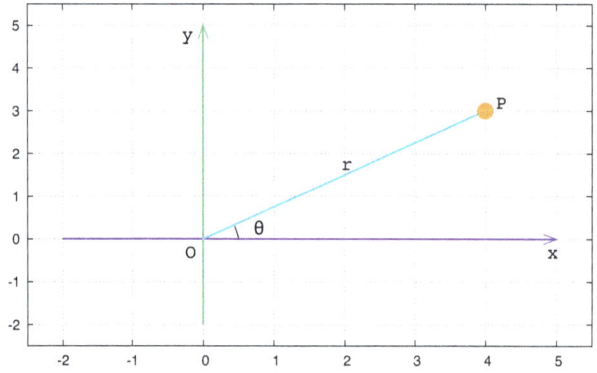

Figure 2.1: Illustration of two-dimensional coordinates

In this example, the point P is located at a values of $x = 4$ along the horizontal from O and a value of $y = 3$ upward. An equivalent location method uses a net distance r and some convenient angle, θ (Greek 'theta') in this case measured anti-clockwise relative to the x-axis.

And so on for higher dimensions[2]. Groupings of coordinates can form a unitary quantity called a *vector*, so the above point P can be referenced by the numeric pair (4, 3); see Chapter 14.

In case you wondered, the polar r-θ coordinates have been used in creating the lemniscate-like image used to brighten up otherwise blank pages; specifically, the shape is based on the equation $r = \alpha \cos^2(2\theta)$, with α setting the overall size.

---

[2]In four-dimensional space-time, one of the coordinates has a different character (see Chapter 29). And string theory needs ten or eleven dimensions, most of which are too compact to be observed (we don't go into that).

Chapter 3

TRIGONOMETRY

THIS is probably the subset of mathematics that initially sparked my interest in the subject[1]. Suppose we want to measure the height of a fairly tall tree in the back garden, perhaps because of the need to fell it, while hopefully not demolishing the neighbour's fence.

If the tree is more or less upright and the surrounding ground is basically flat, the situation can be idealised in the form of a right-angled triangle, as shown in Figure 3.1. The upright bit represents the tree and the horizontal line stands for the ground.

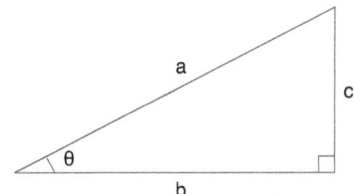

Figure 3.1: A right-angled triangle

The lengths of the sides are marked by the letters a, b and c, while the angle θ (Greek 'theta') is the angle between the lines a and b. The small square at the bottom-right corner of the triangle indicates that this angle is 90 degrees (usually written $90°$) — a right-angle: one-quarter of a complete circle when measured from the circle centre.

The ancient Greeks were probably the first to study planar[2] triangles in a systematic manner, and they came to appreciate that triangles of the same shape but different sizes all share the same property of *proportionality*. In effect, once we know the properties of a triangle of one shape and size, we can work out the properties of any other triangle of the same shape but potentially greatly

---

[1] Being poor at most other subjects no doubt also had something to do with it.
[2] Applicable to flat surfaces.

different size. Thus, the ratio c/a is called the *sine* of the angle θ, the ratio b/a is called the *cosine* of the angle and the ratio c/b is termed the *tangent* of the angle. In equation form, these are written:

$$\frac{c}{a} = \sin\theta,$$
$$\frac{b}{a} = \cos\theta, \qquad (3.1)$$
$$\frac{c}{b} = \tan\theta.$$

Which means that $\tan\theta = \sin\theta/\cos\theta$. This can be seen by expanding out and including all of the steps, as follows:

$$\tan\theta = \frac{c}{b},$$
$$= \frac{c}{a}\frac{a}{b}, \text{ the factor } a \text{ cancels, so the right-hand side is unchanged,}$$
$$= \sin\theta\,\frac{a}{b}, \text{ using the above definition of } \sin\theta.$$

But if $b/a = \cos\theta$, then a/b must equal $1/\cos\theta$, and so $\tan\theta = \sin\theta/\cos\theta$.

Now back to the tree height problem. If we can measure b along the ground and gain some idea of the angle θ, we can determine the height c from the last of these equations:

$$c = b\tan\theta.$$

As a schoolboy, I constructed a very basic device for measuring angles using a protractor (see Figure 3.2), plus a short length of thin brass tubing to look through, the tube mounted on a piece of wood pivoted at the red dot and able to rotate through the set of angles. With this device, I had a brief period going round the neighbourhood measuring the heights of just about anything that didn't move[3].

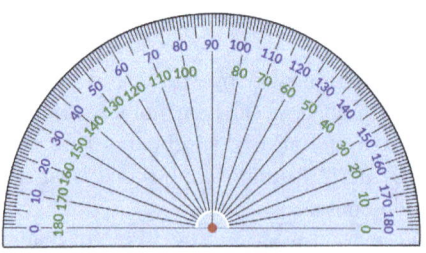

Figure 3.2: A protractor

---

[3]I am reminded of an anecdote related by my father from when he was a schoolboy in Plymouth in the 1920s. The mathematics class had been sent down to Plymouth Hoe to measure the height of Smeaton Tower using trigonometry. Perhaps inevitably, they had paid little attention in the lesson, so one boy was sent off into town to buy a large ball of string ... What the mathematics master later said was unfortunately not recorded.

The basic idea is similar to the *astrolabes* used by early astronomers, and in a more developed form becomes the sextant used in ship navigation (at least until until the advent of GPS). The graduated scales shown in Figure 3.2 are marked in degrees, with 360 degrees (hereafter abbreviated to 360°) in a full circle. The other main angular measure used is *radians*, with 2π radians forming a complete circle.

Given b and the angle θ, all that is then needed is a scientific calculator or a table of tangent values (or a slide rule, for those of a certain age group), plus a bit of multiplication. For example, if the length b is measured to be 20 metres and the angle is 30° (or $\pi/6$ in radians), then the height $c = 11.547$ metres to three decimal places.

Although it is not obvious, the angles in a triangle must sum to 180°, or (equivalently) π radians. Why should this be so? Refer to Figure 3.3:

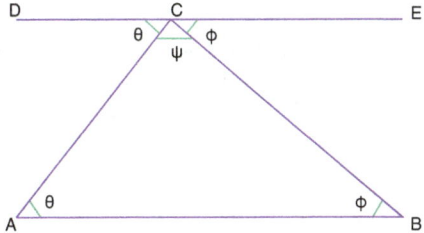

Figure 3.3: Two parallel lines and a triangle

If we imagine drawing a complete *anti*-clockwise circle starting from point E, passing through point D and returning to E, we will have gone through a total angle of 360°, or 2π radians (see Chapter 3). Therefore, going only half-way round must contain an angle of 180°, or π radians. So, describing half a circle anti-clockwise from point D, past C, to E corresponds also to 180° — in which case the three angles marked near C must sum to 180°.

But lines DCE and AB are constructed to be parallel, so the angle $\widehat{DCA}$ must — from symmetry — be equal to angle $\widehat{CAB}$, both then being θ as marked in Figure 3.3[4]. In a similar manner, angle $\widehat{ECB}$ equals angle $\widehat{ABC}$ equals ϕ.

So the sum of the three angles in a triangle, $\theta + \phi + \psi = 180°$. This is valid for planar (flat, Euclidean) geometry; triangles drawn on curved surfaces follow different rules [7].

Before moving on to a slightly more complex trigonometric application, we can use the theorem of Pythagoras[5] to derive a useful property of the sine and cosine functions. The Pythagoras theorem states that for the triangle in Figure 3.1, we must have

$$a^2 = b^2 + c^2,$$

---

[4] These are also called *alternate interior angles* [7].
[5] C. 580 to c. 500 BC.

where the superscript ^{2} implies 'squared' — so that $a^2 = a \times a$. This theorem appears so often in various places and guises throughout this book that a proof (one of several that exist) is provided in Appendix A.

As an aside, this is a convenient place to introduce the $\sqrt{}$ symbol, which is a sufficiently ubiquitous operation as to be added to many hand calculators. If we have $y = x^2$, then $x = \pm\sqrt{y}$ (the plus-or-minus sign $\pm$ appears here since $x^2 = (-x)^2$).

Return now to equation set (3.1), write the first two in the form $c = a\sin\theta$, $b = a\cos\theta$, and substitute into Pythagoras' theorem:

$$a^2 = a^2\cos^2\theta + a^2\sin^2\theta,$$
$$= a^2\left(\cos^2\theta + \sin^2\theta\right),$$

from which it may be deduced that $\cos^2\theta + \sin^2\theta = 1$.

The notation $\cos^2\theta$ stands for $(\cos\theta)^2$; this avoids confusion with $\cos\theta^2$ which would normally be interpreted as $\cos(\theta^2)$.

Suppose now that we cannot access the base of the tree to measure the distance b, perhaps due to the tree being in a lion enclosure, but we can measure the distance d in Figure 3.4 as well as the second angle ϕ (Greek 'phi'). There are then two right-angled triangles, one formed by sides a, $d+e$ and c, and one formed by sides f, e and c.

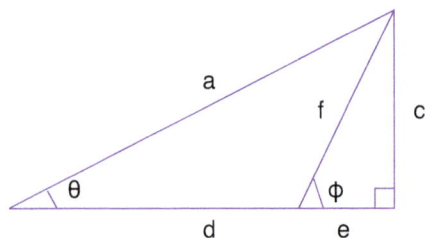

Figure 3.4: More complex height measurement

Using only the tangents of θ and ϕ, on the basis of Figure 3.4:

$$\frac{c}{d+e} = \tan\theta, \tag{3.2}$$

$$\frac{c}{e} = \tan\phi. \tag{3.3}$$

Of these two equations, the quantities d, θ and ϕ can be measured, while c and e are 'unknowns' and c is what we are really interested in. To obtain c, a certain amount of algebra is needed, and the various steps are laid out in some detail[6] as follows:

---

[6] This is the only place where I do show every algebraic step involved. Otherwise the book would balloon to the point where the wood could not be seen for the trees (near literally so for a print edition).

1. Use equation (3.3) to obtain e in terms of c and $\tan\phi$. That is, multiply both sides by e to get $c = e\tan\phi$ and then divide both sides by $\tan\phi$, resulting in:
$$e = \frac{c}{\tan\phi}. \tag{3.4}$$

2. Multiply both sides of equation (3.2) by $d+e$, to get:
$$c = (d+e)\tan\theta,$$

3. and expand out using the distributive law:
$$c = d\tan\theta + e\tan\theta.$$

4. Substitute for e from equation (3.4), so that:
$$c = d\tan\theta + \frac{c}{\tan\phi}\tan\theta.$$

5. Subtract $c\tan\theta/\tan\phi$ from both sides:
$$c - \frac{c\tan\theta}{\tan\phi} = d\tan\theta.$$

6. Multiply both sides by $\tan\phi$:
$$c\tan\phi - c\tan\theta = d\tan\theta\tan\phi.$$

7. Separate out the common factor c on the left-hand-side:
$$c(\tan\phi - \tan\theta) = d\tan\theta\tan\phi.$$

8. Divide both sides by $\tan\phi - \tan\theta$ (assumed to be non-zero) to finally end up with:
$$c = \frac{d\tan\theta\tan\phi}{\tan\phi - \tan\theta}.$$

So we need only measure one length and two angles to obtain the height c, despite not being able to measure e directly. Should e be required also, it can be calculated from equation (3.2) once c is known.

With a bit of practice, algebraic manipulation such as is in the list above becomes second nature, and much of it can be done in one's head.

Two identities regarding sines and cosines will be useful later on, and may as well be inserted here:
$$\sin(\theta+\phi) = \sin\theta\cos\phi + \cos\theta\sin\phi, \tag{3.5}$$
$$\cos(\theta+\phi) = \cos\theta\cos\phi - \sin\theta\sin\phi. \tag{3.6}$$

These can be derived from a pair of compounded right-angle triangles [7], one on top of the other, as shown in Figure 3.5.

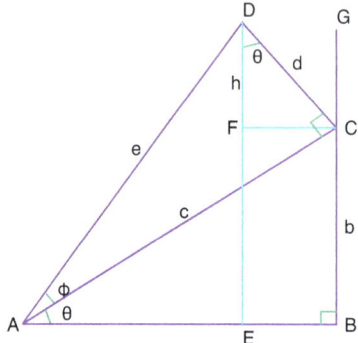

Figure 3.5: Stacked right-angle triangles

The corners of the triangles, as well as the point G, are marked by upper-case letters and the side lengths by lower-case ones. What we are after is the sine and cosine of $\theta + \phi$; look at the first of these:

$$\sin(\theta + \phi) = \frac{DE}{e}; \text{ the ratio of the length } DE \text{ to the length } DA.$$

But $DE = b + h$, where h is the distance from point F to D and F is at the same height as C. Therefore,

$$\sin(\theta + \phi) = \frac{b+h}{e} = \frac{b}{e} + \frac{h}{e},$$
$$= \frac{b\,c}{c\,e} + \frac{h}{e}, \text{ (cancel the } c \text{ to see that this is the same)},$$
$$= \sin\theta \cos\phi + \frac{h}{e},$$

using the definition of $\sin\theta$ for triangle ABC and the definition of $\cos\phi$ for triangle ACD.

To work out the ratio h/e, we need to show that the angle $\widehat{EDC}$ is the same as angle $\widehat{BAC}$, as marked on Figure 3.5. But angle $\widehat{ACB} = 90° - \theta$, since the three angles of this triangle must sum to $180°$. So,

$$\text{angle } \widehat{BCG} = 180° \text{ must also equal } 90° - \theta + \widehat{ACG},$$

in which case angle $\widehat{ACG} = 90° + \theta$, and angle $\widehat{DCG} = \theta$. Since, then, lines BG and ED are parallel, angle $\widehat{EDC} = \theta$.

We thus have,

$$\frac{h}{e} = \frac{h}{d}\frac{d}{e} = \cos\theta \sin\phi,$$

using the definition of $\cos\theta$ for triangle FDC and the definition of $\sin\phi$ for triangle DAC, and so end up with

$$\sin(\theta + \phi) = \sin\theta\cos\phi + \cos\theta\sin\phi.$$

The other equation for $\cos(\theta + \phi)$ follows similar reasoning and is left as an exercise.

Equations (3.5) and (3.6) form the basis (by setting $\phi = \theta$) for the following double-angle formulas, which will come in handy later on:

$$\sin 2\theta = 2\sin\theta\cos\theta, \tag{3.7}$$

$$\cos 2\theta = \cos^2\theta - \sin^2\theta. \tag{3.8}$$

Another relation between the sides of a triangle and their opposite angles is the well-known sine rule equation:

$$\frac{a}{\sin A} = \frac{b}{\sin B} = \frac{c}{\sin C}, \tag{3.9}$$

with reference to Figure 3.6. The labels given to sides and angles helps in memorising the equation: angle A is opposite to side a, and so on.

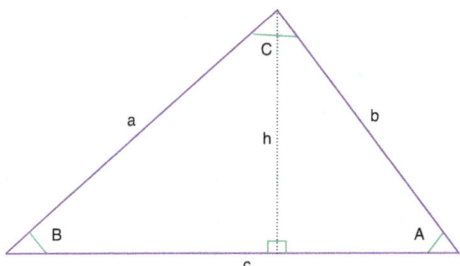

Figure 3.6: Triangle for the sine rule

As to where equation (3.9) comes from, just use the intermediate length h dropped from the top of the triangle down to the base on side c. Then, for angles A and B, using the two right-angled triangles and the analogue of equation (3.1):

$$\frac{h}{a} = \sin B \text{ and } \frac{h}{b} = \sin A.$$

Rearranging this then results in

$$a\sin B = b\sin A, \text{ or } \frac{a}{\sin A} = \frac{b}{\sin B}.$$

The third element of equation (3.9) follows in a similar manner, but now bisect angle A (or angle B) rather than angle C and so draw the perpendicular line onto side a.

Equation 3.9 has found a ready use in surveying, since it is only necessary to start with one accurately measured baseline length[7], c say; thereafter only

---

[7]Not always an easy task to carry out in difficult terrain [8].

angles need to be measured. This enables a network of linked triangles to march across the landscape, as illustrated by Figure 3.7, thus building up a map of the terrain in terms of distances between distinctive points[8].

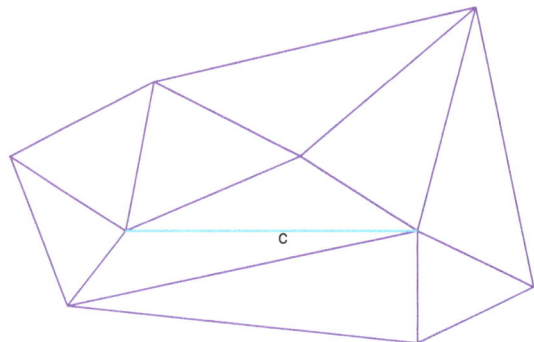

Figure 3.7: Surveying flat terrain

In practice, some adjustment of the geometry is needed to accommodate changes in altitude, but equation (3.1) helps here also.

The mapping of larger areas of the earth's surface requires consideration of spherical trigonometry, thus taking account of the surface curvature. This goes a bit beyond the material I can easily explain in the present book, though.

Figure 3.6 can also be used to derive the cosine rule, which has the form:

$$b^2 = a^2 + c^2 - 2ac \cos B,$$

which can be regarded as a generalisation of the Pythagoras equation to a planar triangle of any shape. The derivation is straightforward, again using the vertical distance h as an intermediary quantity, and here subdividing side c into two quantities either side of the vertical, with d on the left and e on the right, so that $c = d + e$. Then use the following:

$$b^2 = h^2 + e^2, \text{ from the right-angled triangle on the right,}$$
$$a^2 = h^2 + d^2, \text{ from the right-angled triangle on the left.}$$

Subtract these and rearrange to get $b^2 = a^2 - d^2 + e^2$. Use $d + e = c$ to obtain $e = c - d$ and substitute this in to form

$$b^2 = a^2 - d^2 + (c - d)^2 = a^2 + c^2 - 2cd.$$

Finally use the definition of $\cos B = d/a$ from the left-hand right-angled triangle to express d in terms of a, c and $\cos B$ and we arrive at the cosine rule. Similar equations obtain for the other two sides of the triangle.

---

[8]In the UK, these are 'trig points': concrete pillars designed as mount points for theodolites.

Chapter 4

ALGEBRAIC EXPRESSIONS

TWO examples here show what can be achieved using a basic amount of 'problem modelling', combined with some straightforward algebra.

4.1 Pension Drawdown

Suppose that we have reached retirement and need to decide how best to make use of our accumulated pension fund[1]. One of the options now available to retirees (at least in the UK) is 'income drawdown', where sums of money can be taken per month or per year or whatever, and where the amounts can be varied at any time.

Assume at the outset that the size of the pension fund is P_0 (in £ sterling), that the pension administration company charges c percent of the residual fund per annum and that the stock market (or the relevant invested funds) grow at g percent per annum on average. In actuality, g will vary more or less randomly from year to year (or even day to day), but for present purposes a fixed value will suffice. Suppose that it is desired to extract y (also in £) per annum, to live on.

Introduce a year counter k, commencing from zero, such that the value of the fund at the outset of any year k is P_k. At the very next year, the fund value will then be:

$$P_{k+1} = P_k - 0.01cP_k + 0.01gP_k - y. \tag{4.1}$$

That is, the fund has been diminished by the administration charges, augmented (hopefully) by stock market growth, and decreased again by the income drawdown amount y. The factors 0.01 occurring here act to convert the percentage quantities into proportions.

---
[1] Why it is usually termed pension 'pot' is unclear; maybe in days past it was stored in a pot to be buried.

Even as it stands, equation (4.1) is informative; with a pocket calculator or a spreadsheet we can work several years into the future with different values of y and best-estimates for c and g, and see what can be afforded and for how long. By way of example, suppose $P_0 = £250{,}000$, $c = 0.75\%$ and $g = 5\%$. Using an assortment of values of y and graphing the results generates Figure 4.1.

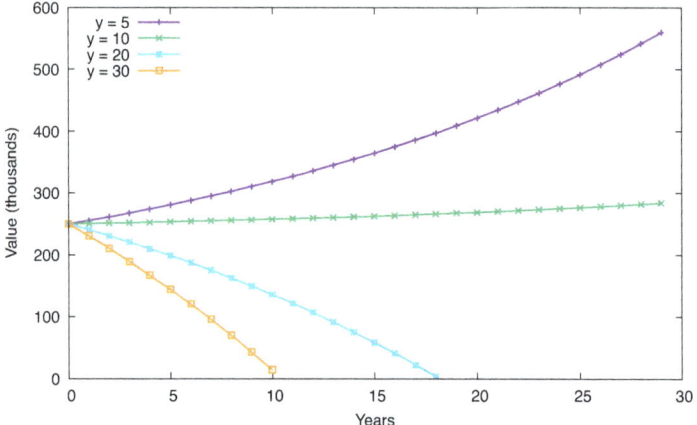

Figure 4.1: Examples of pension drawdown

The monetary values are in thousands. It can be seen that if we do not extract too much, the fund can actually continue to grow, provided the stock market behaves as we hope.

With a certain amount of algebra, however, we can simplify the extraction of more information — in particular, what amount y can be withdrawn per year without altering the value, plus a more direct means for determining when the money will run out. To do this, write equation (4.1) in the form

$$P_{k+1} = \alpha P_k - y,$$

which uses the distributive law in Section 2 to separate out the quantity $\alpha = 1 - 0.01c + 0.01g$, assumed constant, from y.

This is termed a *first-order difference equation*, which can be used to infer P_k directly, for any k. To see this, build up the sequence, commencing from zero:

$$P_1 = \alpha P_0 - y,$$
$$P_2 = \alpha P_1 - y = \alpha^2 P_0 - (1 + \alpha) y,$$
$$P_3 = \alpha P_2 - y = \alpha^3 P_0 - (1 + \alpha + \alpha^2) y,$$

and so on.

Using the pattern visible here, we can infer that

$$P_k = \alpha^k P_0 - \left(1 + \alpha + \ldots + \alpha^{k-1}\right) y, \qquad (4.2)$$

4.1. PENSION DRAWDOWN

for any k greater than zero.

The quantity α^k stands for α multiplied by itself k times: $\overset{\underset{\longleftarrow}{k}\underset{\longrightarrow}{}}{\alpha \times \alpha \times \ldots \times \alpha}$. Also, the '...' construct above is simply shorthand for absent terms in the series; there is no need to write absolutely everything if the reader can infer the missing bits.

Equation (4.2) is still not quite in the most convenient form, but the series $1 + \alpha + \ldots + \alpha^{k-1}$ is what is termed a *geometric series* and, with reference to [9], we find that

$$1 + \alpha + \ldots + \alpha^{k-1} = \frac{1 - \alpha^k}{1 - \alpha},$$

provided $\alpha \neq 1$. Therefore,

$$P_k = \alpha^k P_0 - \left(\frac{1 - \alpha^k}{1 - \alpha}\right) y. \tag{4.3}$$

This simplifies the task of answering the questions of interest, which are examined further below.

4.1.1 Keeping the Balance the Same

Suppose that we need to choose a drawdown value y that is just large enough to cancel the net growth in the fund (assuming this growth is positive), so that $P_k = P_0$ for any value of k greater than zero. In effect, the pension pot remains at the same value, year by year. Substituting this into equation (4.3) and rearranging the terms results in:

$$\left(1 - \alpha^k\right) P_0 = \left(\frac{1 - \alpha^k}{1 - \alpha}\right) y.$$

Provided $1 - \alpha^k$ is not zero (which would require $\alpha = 1$ or $k = 0$), this common term can be cancelled out, resulting in

$$y = (1 - \alpha) P_0,$$

this being the drawdown amount that will leave the fund value unchanged from year to year. It may be inferred that α needs to be less than one if withdrawals are to make sense; otherwise, with $\alpha > 1$, we would instead need to put money in (since the stock market growth is not enough to overcome the administration charges).

4.1.2 When Does the Money Run Out?

To see how long the fund will last under current conditions, assuming the same y deduction every year, simply set equation (4.3) to zero:

$$\alpha^k P_0 - \left(\frac{1 - \alpha^k}{1 - \alpha}\right) y = 0.$$

Multiply both sides by $1 - \alpha$ (assumed non-zero), to get:

$(1 - \alpha)\alpha^k P_0 - (1 - \alpha^k) y = 0$, which may be written $(1 - \alpha)\alpha^k P_0 = (1 - \alpha^k) y$.

Take the α^k term on the right over to the left to join its sibling (use the distributive law again):

$$\alpha^k \left\{ (1 - \alpha) P_0 + y \right\} = y,$$

which finally gives

$$\alpha^k = \frac{y}{y + (1 - \alpha) P_0}. \tag{4.4}$$

To obtain an explicit expression for k, the *logarithmic*, or log, function is needed. The two most common varieties are log to base 10 and log to base e, both of which are inverses of a power function and both of which are typically available on a scientific calculator. That is,

$$y = 10^x \Leftrightarrow x = \log_{10} y,$$
$$y = e^x \Leftrightarrow x = \log_e y,$$

the symbol $\Leftrightarrow$ standing for 'implies and is implied by'. The quantity $e = 2.7182818...$ is another irrational number which crops up all over the place in mathematics; it will be discussed further in Chapter 7.

So take the log to base e of equation (4.4), to obtain[2]

$$k = \frac{1}{\log_e \alpha} \log_e \left\{ \frac{y}{y + (1 - \alpha) P_0} \right\}.$$

This provides a more direct way of working out how long the money will last, albeit with a little work using a calculator, and maybe a bit of paper to retain intermediate results.

4.2 Recursive Average

For a short while, I worked for a company that provided marine and aerodynamic testing, using water tanks for the former and wind tunnels for the latter. In both cases, characterisation of the flow field around the object (ship, aerofoil or whatever) was obtained using a distributed set of probes, each probe connected to an analogue-to-digital converter and the resulting digital data stored on a computer. Each such data stream was then averaged to obtain a representative value of pressure, temperature or whatever, at that point around the object. The nature of turbulent flow, in particular, meant that quite astonishing amounts of data needed to be stored in order that sufficiently accurate averages could be

---

[2] Log to base e is preferable in many cases, since the exponential function has unique properties. This will become apparent in Section 7.2.

4.2. RECURSIVE AVERAGE

obtained; and this was in the days when computer memory was at something of a premium.

So it would be nice if the average can be worked out as we are going along, preferably with some associated measure of uncertainty, so that the data collection can be terminated once sufficient certainty is achieved. This is possible using what is termed a *recursive average* formula, and the only storage needed is for the current average and the number of samples collected thus far.

Firstly, some notation: denote the data sequence to be averaged as the k-valued set $x_1, x_2, x_3, \ldots, x_k$, for a total of k values. Then define the average of these items in the usual way as the sum of all of them divided by k, namely:

$$y_k = \frac{1}{k} \sum_{j=1}^{k} x_j. \tag{4.5}$$

This is just shorthand for $x_1 + x_2 + \ldots + x_k$, the whole sum then divided by k. The symbol $\sum$ is the capital Greek letter 'sigma' (the lower case letter being σ), and stands for 'sum' over the discrete set of values. The symbol j is employed as a variable index, or counter, here running from 1 to k inclusive (the notation $\sum_{j=1}^{k}$ specifies that the sum is constrained between lower and upper bounds of $j = 1$ and $j = k$ respectively, with the upper limit abbreviated here to just k since the meaning is clear).

Now for a bit of algebra: write

$$y_k = \frac{1}{k} \sum_{j=1}^{k-1} x_j + \frac{1}{k} x_k, \tag{4.6}$$

which just extracts the last term in the average. But, from the definition of the average and with reference to equation (4.5),

$$y_{k-1} = \frac{1}{(k-1)} \sum_{j=1}^{k-1} x_j,$$

which is just the average over $k - 1$ values. Multiply both sides by $k - 1$ and rearrange:

$$\sum_{j=1}^{k-1} x_j = (k-1) y_{k-1}.$$

Then use this equality to replace the summation term $\sum x_j$ in equation (4.6) with $(k - 1) y_{k-1}$. Therefore,

$$y_k = \left(\frac{k-1}{k}\right) y_{k-1} + \frac{1}{k} x_k,$$

which is the result sought.

It is possible to obtain a similar expression for the data *variance*, denoted σ_k^2 and defined as:

$$\sigma_k^2 = \frac{1}{k} \sum_{j=1}^{k} (x_j - y_k)^2.$$

This is the square of the standard deviation σ_k and a measure of how variable is the x-data stream about its mean value y_k. The algebra is a bit more convoluted, however, and the recursive equation is simply stated here:

$$\sigma_k^2 = \left(\frac{k-1}{k}\right) \sigma_{k-1}^2 + \frac{1}{k-1} (y_k - x_k)^2, \quad \text{valid for } k > 1.$$

Of perhaps more interest is some idea of how certain is y_k, since y_k itself will vary with k. Thus, the standard deviation of y_k is given by

$$\frac{\sigma_k}{\sqrt{k}}. \tag{4.7}$$

This is the standard deviation σ_k associated with the x-data stream, divided by the square root of the number of data samples. Deriving this equation requires Kalman Filter-type analysis or similar, which is touched on in Chapter 21. The certainty of y improves rather slowly as k increases, such behaviour being characteristic of sampling in general (as in polling people's opinions on some topic or other, k then standing for the sample size).

As an illustration, suppose that a set of random data points is generated, with a defined mean value of 2 and known standard deviation 1, and assume that $k = 200$. Figure 4.2 then illustrates the above findings as k increases.

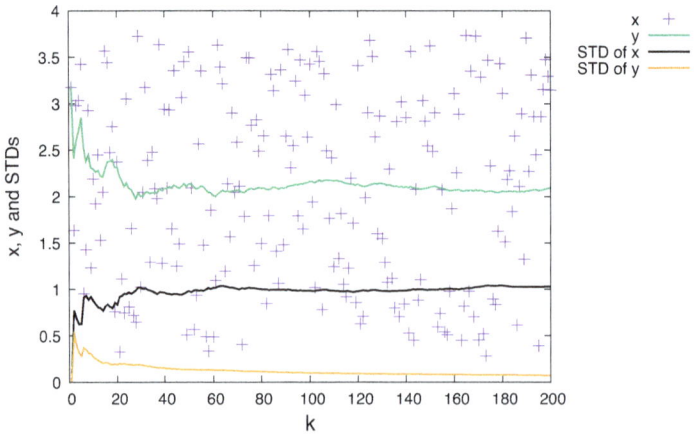

Figure 4.2: Examples of x_k, y_k and their standard deviations

Here,

4.2. RECURSIVE AVERAGE

- The purple crosses denote the random sampled x-values, acting as a surrogate for data that could be obtained as part of some experiment (so for present purposes we need to pretend that we don't know that x has a mean of 2 and standard deviation of 1).

- The green line stands for the recursive average y_k, which can be seen to converge (slowly) to the expected value of 2.

- The light blue line is the recursively-averaged standard deviation of x, converging to unity as expected.

- And the light brown line denotes equation (4.7) for the uncertainty (standard deviation) associated with y_k. Only this curve tends to zero as k increases.

As a final remark here, the random data for x_k was generated using a *uniform distribution*, in which values are equi-probable within a given range. Other random distributions, such as the *normal distribution*, tend to be more or less concentrated. The normal, or Gaussian[3], distribution has the general equation for the probability density (see [6], for example):

$$f(x) = \frac{e^{-(x-x_0)^2/2\sigma^2}}{\sigma\sqrt{2\pi}},$$

and is illustrated in Figure 4.3, here using a mean value $x_0 = 2$ and a standard deviation $\sigma = 0.8$.

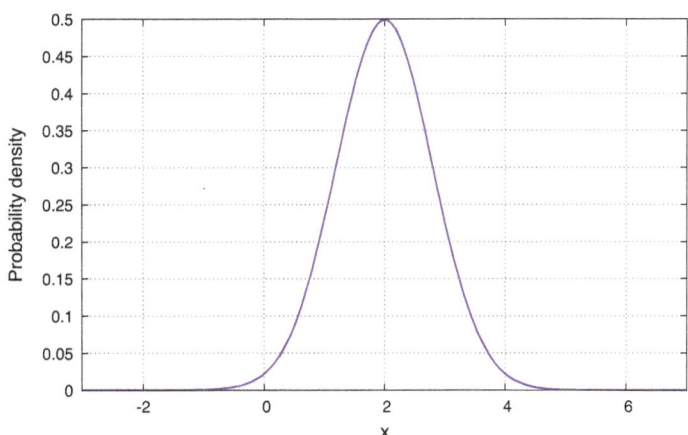

Figure 4.3: Illustration of the normal distribution

For such a continuous distribution, the probability P of finding a value of x in the small region $[x, x+dx]$ is given by $f(x)\,dx$.

---

[3] After Carl Friedrich Gauss, 1777 to 1855.

Chapter 5

SIZE OF A KILLER ASTEROID

IT is now generally accepted that the demise of the dinosaurs (and numerous other species, both terrestrial and marine) 66 million years ago can be attributed to the impact of an asteroid. Up until 1980, numerous mechanisms had been proposed, including an impact, but there was no direct evidence for any of the theories nor any way of assessing their respective viabilities. That situation changed with the publication of a paper by the Alvarez team [10], which not only provided relevant evidence but also attempted to infer the mass and size of the impacting body itself. It is the latter claim which is of interest here, since the equation for the mass is simple but with profound consequences; it is as follows:

$$M = \frac{sA}{0.22f}. \tag{5.1}$$

The various constituents here will be defined in due course.

What the Alvarez team concentrated on was the presence of a puzzling concentration of the rare metal iridium in a thin clay layer from the time of the extinction, this layer separating the Cretaceous and Tertiary periods 66 million years ago[1]. Iridium is very uncommon at the earth surface but known to be present in greater quantities in exo-terrestrial bodies such as asteroids, so it was not a particularly great stretch to attribute the iridium layer to an impact event. Nonetheless, the claim was a bold one at the time, since no impact crater of the right age and size had been identified; that came later.

The logic of equation (5.1) is straightforward: estimate the mass of iridium in the boundary clays, on the assumption that this metal had been distributed more or less uniformly around the globe. Then equate this to the expected mass of iridium present in the asteroid prior to impact, a proportion of latter being then blasted into the upper atmosphere.

---

[1] Known familiarly as the K-T boundary or — more recently — the K-Pg boundary, Pg standing for Paleogene.

The former quantity is given by sA, where s is the iridium concentration in the boundary layer (this can be measured) and A is the surface area of the earth.

The second quantity — the mass of the iridium dispersed in the upper atmosphere — is $0.22fM$, in which f is the expected concentration of iridium in carbonaceous-chondrite asteroids and M is the asteroid mass (the quantity of primary interest here). The factor 0.22 represents the proportion of the disintegrated asteroid that ended up in the upper atmosphere — in the Alvarez paper, in the absence of any other information, this was a figure based on the well-studied eruption of the Krakatoa volcano in 1883[2].

So we get $0.22fM = sA$, which can be rearranged to provide equation (5.1).

It is of interest now to plug some numbers into the above equation, to get an idea of how big the asteroid was. Firstly, though, a brief note on scientific notation: for example, 2×10^{-5} stands for 2 multiplied by 10^{-5}, which in turn is equal to 2 divided by 10^5, and where 10^5 is 10 multiplied by itself 5 times. So $2 \times 10^{-5} = 0.00002$. Similarly, 2×10^6 stands for 2 with six zeroes after it, or 2 million. And so on. For very large or very small numbers, the benefit of such compactness is obvious.

So, we have, in SI units[3] :

- $s = 8 \times 10^{-8}$ kilograms per square metre (units usually abbreviated to kg m^{-2}). This, the amount of iridium in the boundary clays, was based on measurements of rocks at the K-Pg boundary in Italy and Denmark.

- $A = 5.101 \times 10^{14}$ m^2, the surface area of the earth [11].

- $f = 5 \times 10^{-7}$, the mass fractional abundance of iridium in carbonaceous-chondrite asteroids.

Carrying out the calculations, we arrive at a value of $M = 3.71 \times 10^{14}$ kg. This is the asteroid mass, which can then be used to infer the expected energy at impact, in turn allowing an estimation of crater size and — importantly — of the likely resulting damage [12]. From a human perspective, such a mass value may not be especially meaningful, but it can be converted into a size based on some idea of the asteroid density ρ (Greek letter 'rho'), assuming $\rho = 2200$ kg m^{-3} [10].

If the asteroid was assumed to be more or less spherical in shape (unlikely to be exact but the assumption will give useful results), then

$$M = \frac{4}{3}\pi r^3 \rho,$$

with r being the radius. That is, doing the sums, $r = 3427$ m, or a diameter of about 6.8 km. Further estimates in [10] suggest it was about 10 km across.

The impact crater was discovered in 1991 just offshore of the Mexican Yucatán peninsula near a village called Chicxulub. The crater measures about 150 km across, consistent with expectations based on the size of the asteroid [12].

---

[2]Specifically, 18 cubic kilometres of material was ejected into the atmosphere, of which 4 cubic kilometres ended up in the stratosphere, the latter then being distributed around the globe.

[3]Système Internationale, based on metres, kilograms and seconds.

Chapter 6

PROBABILITY

PROBABILITY can be described as dealing with the characterisation of randomness, and has a reputation of being 'difficult'. This description is, perhaps, not entirely unearned, but it would be more accurate to say that many probabilistic predictions are counter-intuitive — going against what we would expect. Partly this is due to the innate propensity for humans to see patterns, even when none are there — for example, Shakespeare's Hamlet seeing images in cloud formations. The example in Section 6.1 below is a prime case of this propensity, and another instance is the apparent alignments in randomly distributed points [13].

Section 6.2 below is now especially pertinent, given the increased use of medical tests as a first line of diagnosis for more and more medical conditions. It is also the case that, as a consequence of the recent coronavirus pandemic, the general public is now much more aware of the limitations of medical tests (of which more below), and this awareness can only be beneficial.

A few words on the combination of probabilities. If two events A and B are known to be independent (and they do need to be independent, rather than just hoped to be so), with associated probabilities p_A and p_B, then the probability of occurrence of A *and* B will be the product $p_A p_B$. Similarly, if the events are mutually exclusive, the probability of occurrence of A *or* B will be the sum $p_A + p_B$ [14].

6.1 Gambler's Ruin

Suppose we are betting on coin tosses (not a good idea, but bear with the supposition), and that 6 heads have occurred in succession, thus:

...THHHHHH.

What should we bet on next? Two common forms of reasoning go as follows:

- "Wow! I'm in a run of heads, so the next coin toss is likely to be heads as well. I will bet on that."

- "Hmm. It's a run of heads, but overall I expect equal numbers of heads and tails to occur, so I'll bet on tails this time round."

In fact, neither line of reasoning is correct. The point about coin tosses[1] is that each toss is independent of the preceding one, and heads and tails can occur with equal probability of $1/2$.

But what about the run of 6 heads — shouldn't that be quite unlikely? Well, no; it turns out that the longer we toss coins and record the results, longer and longer runs of heads and tails will occur. To show this, it is necessary to dip slightly into the maths of probability and make use of the rule for the combination of independent events: so the probability of getting two heads (or tails) in succession is $1/2 \times 1/2 = 1/4$.

To proceed further, concentrate on runs of heads and make a distinction between runs that are *exactly* k long, and runs that are *at least* k long, for some integer k. This distinction brings out an important aspect of probability theory and analysis: the need to be precise in problem formulation.

Let the probability that a head will occur be p, and the probability of a tail be q. Yes, they are the same in the current example, but it helps the analysis to make a distinction, besides generalising the result to situations of non-equal probability. We can then say that the probability of exactly k heads will correspond to the sequence

$$\ldots T \overset{\longleftarrow k \longrightarrow}{H H \ldots H} T \ldots.$$

The logic here is that a run of exactly k heads must start and end with a tail, which act as delimiters. So the corresponding probability will be given by:

$$P_k = q^2 p^k,$$

since there are two tails and k heads.

In a similar manner, the probability of at least k heads will be:

$$P_{k+} = qp^k,$$

since we are not concerned whether the next toss is heads or tails (what comes after is of no concern, since we have accumulated the k heads).

These give the probabilities that specific k-length runs will occur, but it is also instructive to obtain the expected *maximum* length of a run, if we toss the coin a total of n times. Out of these n tosses, we expect nqp^k occurrences of at least k-length runs, while the maximum run length will occur perhaps only once or so. That is,

$$nqp^k = 1,$$

for the maximum length k, or the nearest integer thereto [15].

Putting some numbers into this equation, with $p = q = 1/2$, results in Figure 6.1.

---

[1]'Fair' coins, that is, that have not been tampered with.

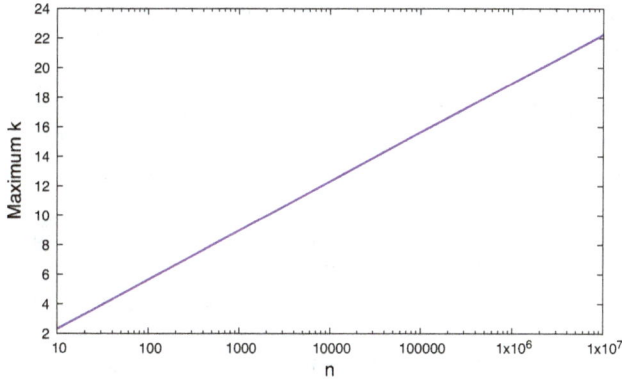

Figure 6.1: Maximum expected k versus n

The values of n along the bottom axis are plotted in logarithmic scale, which exposes the linear growth in maximum run length with $\log_e n$. In summary, if we toss a fair coin long enough, there is no limit to the run lengths that can be observed. Not something that would have been expected.

As an aside here, given a reasonable level of programming experience, it is not difficult to verify the above results using numerical simulations[2].

And as a further aside, the *absence* of runs of the expected length provides a useful way of discerning between 'made-up' coin-toss sequences and the genuine article [16].

6.2 Medical Tests

Suppose the following scenario has occurred: you have gone to your doctor complaining of a persistent pain in some part of your anatomy. The doctor suspects that you might be suffering from disease X (to avoid complaints from the medical profession, I am being deliberately vague here), and orders a particular test to be carried out. The test comes back positive, and when you are called back in to see your doctor, he tells you that this test is 95% accurate. Is this a reason to be worried?

Unfortunately, it is not always clear in the consultation quite what the 95% probability actually refers to, and the usual interpretation is that this is the probability that you have disease X. The above situation is by no means uncommon, since the interpretation of probabilistic values is not something that most general practitioners, or even consultants, are necessarily familiar with (to be fair, they already have enough to worry about). Understanding the results is also important in the context of home testing kits, and recently perhaps particularly relevant to the *lateral flow test* used for detecting the presence of the SARS-CoV-2 antigen [17].

To be better informed as to what is going on, and to determine whether you should be worried or not, it is necessary to deal with *conditional probabilities*.

---

[2]Certainly to be preferred over manual coin tossing.

A conditional probability is the probability that an event will occur, *given that* another event has already happened. There is an implicit ordering of events here. The probability that event A occurs, given that event B has already occurred, is usually written $P(A|B)$, and read as 'probability of A given B'. The following calculations are an application of Bayes' theorem[3] [18], which enables prior estimates of probability to be revised in the light of new observations or information.

It is also necessary to appreciate that any medical test can fail in two distinct ways: it can fail to detect if you have disease X when you actually do (a 'missed detection'); or it can say you have the disease when you genuinely don't (a 'false alarm'). For the test to be useful, both of these failure modes should have small associated probabilities, although a greater number of false alarms can generally be tolerated. There may also be a third failure mode when there is no result at all, but this situation provides no information and can be ignored for present purposes.

A third number is important: the background rate of disease X in the general population, meaning the proportion of the population (or the relevant part of the population[4]) with disease X.

With these three numbers — the background rate and the two failure probabilities — it is possible to make a reasonable stab at determining the actual probability that you have the disease. The easiest way of gaining understanding is to generate a probabilistic tree diagram; this reflects the fact that you are taking part in a medical trial, although without contributing to the trial statistics. First, though, some example numerical assumptions concerning this (mythical) disease X:

- The background rate in the general population = 1%, meaning a 0.01 probability that a person chosen at random will have disease X.

- The test is 95% accurate at detecting real cases of X, meaning a probability of 0.95 correct detection, given that the person is suffering from X.

- The test has a 3% false alarm rate, so the false alarm probability is 0.03.

Then choose a random sample of 10,000 people from the general population and subdivide into two categories: 'X' for disease X and 'H' for healthy. Given the above probabilities, we can expect 100 people to have X and the remaining 9900 to be healthy. Each of these groups is then hypothetically subjected to the medical test in question. Denote a positive test with a + sign and a negative test with a − sign. Tracing through the various proportions at each stage then gives Figure 6.2.

---

[3]Thomas Bayes, 1701/2 to 1761.
[4]For instance, a disease affecting males only.

6.2. MEDICAL TESTS

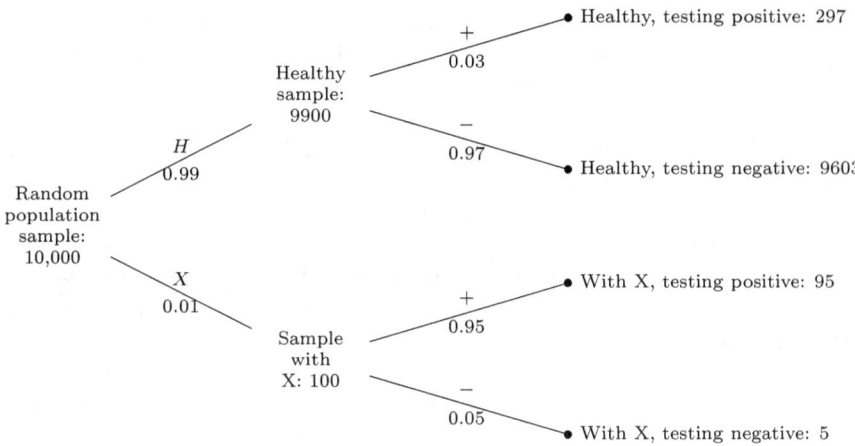

Figure 6.2: Probability tree diagram, in terms of numbers and proportions

The final numbers on the right-hand-side are for each of the four categories. Now, so far as you are concerned, all you know (apart from the initial symptoms) are that you tested positive. Therefore, you could be in either the group 'healthy and positive' or the group 'X and positive' — and the former group is larger.

In fact, the quantity you really want is the probability that you have disease X given a positive test result, which is given by

$$P(X|+) = \frac{95}{95 + 297} = 0.24, \text{ to 2 decimal places.} \tag{6.1}$$

This is a whole lot smaller than the probability that you test positive given X, which is $P(+|X) = 0.95$.

The first takeaway point from this exercise is that the number of false positives will, in general, tend to be larger than the number of real positives, simply due to the low background rate of disease X in the general population. The exception to this statement is if the false alarm rate is extremely low. In any case, it pays to work through the numbers.

The second takeaway point is that $P(X|+) \neq P(+|X)$ — the two quantities are seldom equal. Unfortunately, the assumption of equality is often made in legal cases, with potentially tragic results; see Section 6.3 below.

For completeness, equation (6.1) should be written in symbolic form as:

$$P(X|+) = \frac{P(+|X)P(X)}{P(+|X)P(X) + P(+|H)P(H)}, \tag{6.2}$$

which is one form of Bayes' theorem[5]. Insert the values $P(+|X) = 0.95$, $P(X) = 0.01$, $P(+|H) = 0.03$ and $P(H) = 1 - P(X) = 0.99$ to cross-check. In words, equation (6.2) can be obtained by multiplying the constituent probabilities along

---

[5]I find it easier to remember the simpler form $P(X|+)P(+) = P(+|X)P(X)$, where $P(+)$ is the denominator (the stuff below the line) in equation (6.2).

each of the individual branches in Figure 6.2, and then dividing the 'X' and '+' product value by the sum of the 'X' and '+' and 'H' and '+' product values.

For interest, the complete set of probabilities for the above example are listed in Appendix B.

Now to another aspect of this example. So far as I can determine, a positive test result is usually a signal for some more invasive procedure to take place, such as a biopsy. What seldom seems to happen is a recommendation for a second independent test (and this does mean completely independent, not a re-examination of the original blood sample, or whatever). And this is a bit strange; since the outcome of a test is to some extent random, wouldn't it make sense to repeat it? In the wider sense, we are dealing with *data fusion*, in which newer information can be used to strengthen the evidence, one way or the other. This can be done here simply by extending the above tree diagram to accommodate another layer; the annotation has been abbreviated for clarity.

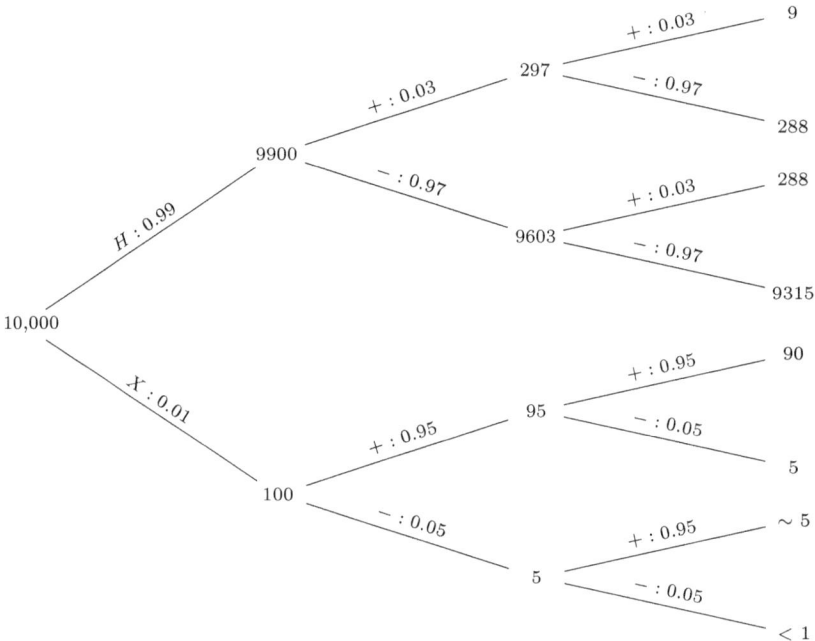

Figure 6.3: Probability tree diagram, with three levels

Numbers have been rounded to the nearest integer and the second test is assumed to have the same associated probabilities as the first one.

Now suppose the second test also came back positive, so examine the branches containing two 'positives'. On the H branch there are 9 cases and on the X branch there are 90. Therefore,

$$P(X|++) = \frac{90}{90+9} = 0.909, \text{ (using the above rounded integers)}, \quad (6.3)$$

which greatly increases the confidence that you do, indeed, have disease X.

In contrast, if the second test came back *negative*, then $P(X|+-) = 0.017$, meaning that you can likely stop worrying (at least so far as disease X is concerned).

Using the probability symbology, equation (6.3) becomes:

$$P(X|++) = \frac{P(+|X)^2 P(X)}{P(+|X)^2 P(X) + P(+|H)^2 P(H)}, \tag{6.4}$$

which is straightforwardly obtained by multiplying the probability values down the relevant branches. It may be observed, though, that this is not quite in the same form as equation (6.2), as those pesky squared quantities upset the pattern. However, equation (6.4) can also be written in the form:

$$P(X|++) = \frac{P(+|X)P(X|+)}{P(+|X)P(X|+) + P(+|H)P(H|+)}. \tag{6.5}$$

This is now of Bayesian form and recursive, in that the existing value of $P(X|+)$ can be updated directly with the new test probability. And $P(X|+++)$, etc, follow the same pattern.

To see that equations (6.4) and (6.5) are actually the same (I agree that it's not exactly obvious), use equation (6.2) to substitute for $P(X|+)$ in equation (6.5). A bit of straightforward algebra, combined with $P(H|+) = 1 - P(X|+)$ (see Appendix B for this) then delivers equation (6.4).

Equation (6.5) is a form of data fusion and Section 21 later on in the book picks up this thread again, albeit in terms of probability distributions or their statistical measures.

6.3 Prosecutor's Fallacy

It would be remiss of me if no mention was made of the sometimes serious miscarriages of justice that have occurred due to the misuse of probability in legal cases. In the UK, the Sally Clark case [19],[20] comes to mind, where estimated probability values were incorrectly multiplied[6] together to give a tiny — and wholly misleading — probability that multiple events could have occurred randomly.

The *prosecutor's fallacy*, or *inversion fallacy*, occurs when the probability of innocence given the evidence is wrongly assumed to be equal to an infinitesimally small probability that that evidence would occur if the defendant was innocent[7] [21] — in effect, confusing $P(I|E)$ with $P(E|I)$, here using I to stand for 'innocence' and E for 'evidence'. The increasing use of DNA evidence in court cases has recently led to concentrated attention on the inversion fallacy, due at least in part to the tiny probability values that are often quoted and to the potential for misunderstandings. This book is not the right place to go into a lot of detail about the derivation of DNA evidence and how it is used

---

[6]It is only valid to multiply event probabilities if they are known to be independent.

[7]The difficulty in untangling this sentence may be contrasted with the clarity of a mathematical equation.

in legal trials, but the figure that is frequently presented in court is the *Random Match Probability* (RMP). This is the theoretical probability that some randomly-chosen person out of the relevant population will match the DNA sample taken from the crime scene: namely, $P(E|I)$. 'Theoretical', because the statistically-derived probability takes no account of errors such as contamination, mislabelling and the various other mix-ups that could occur during the collection and processing of the sample. The actual RMP may be rather larger than the one-in-millions or billions or trillions that tend to be quoted [22] — such tiny figures can result in all the other evidence seeming to be irrelevant. More importantly in the present context, though, the RMP is frequently misinterpreted as the probability that the defendant would be innocent given the DNA evidence: as stated above, incorrectly equating $P(I|E)$ with $P(E|I)$.

To be fair, the fallacy is a subtle error requiring thought to appreciate, but even so I cannot make up my mind whether judges and lawyers are genuinely ignorant of probability theory, or are exploiting the likely ignorance of jurors. Either way, the increasing use of DNA tests in court cases means that jurors need to be aware of the above arguments, should not be entirely reliant on the confidence of expert witnesses, and need to take account of the other evidence presented in the case. The relevance of that other evidence can be appreciated by looking back at Section 6.2, in which equation (6.2) required two other probability values in order to complete the calculation. In the present case, though, those values are either absent or much less well-defined, rendering the calculation of $P(I|E)$ too uncertain to be useful.

For further reading on the use (and misuse) of probabilities in legal cases, see [23], which describes a number of real-life examples. A comprehensive overview of the calculation of DNA match probabilities and their presentation and treatment in US court cases may be found in [24].

6.4 Mark and Recapture

A technique used by biologists to estimate the size of wildlife populations also relies on probability. This is the 'mark and recapture' method (also known as the *Lincoln-Petersen* method) and can be applied to estimating how many fish are in a lake, for example. It proceeds as follows:

- Capture n fish from the lake, mark all of them and release back into the water.

- Some while later on, recapture another set of K fish from the same lake and count the number k that are marked.

- Then the total number of fish N in the lake as a whole can be estimated from:
$$N \approx \frac{nK}{k}.$$

6.5. PROBABILITY OF A NUCLEAR DISASTER

The logic here is that the proportion of marked fish in the second sample, k/K, will be approximately the same as the proportion of the original captured set to the total in the lake, namely n/N.

It is fun to simulate this process, assuming that there are 417 fish total. Then allow for multiple capture-mark-recapture trials, on each trial varying the sample sizes n and K randomly in the range 20 to 60 (entirely arbitrary limits). The results of 500 such trials is plotted in Figure 6.4.

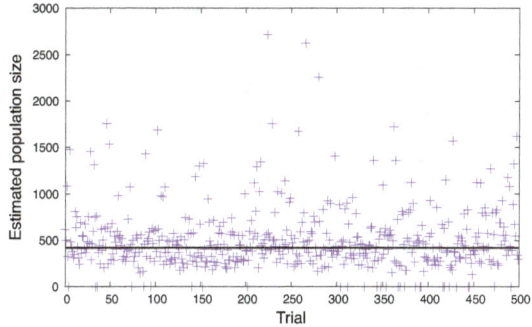

Figure 6.4: Mark and recapture simulation

The horizontal black line marks the true population size, and it can be seen that the estimated values are for the most part quite close to this. The method is random, however, and success is not guaranteed — small values of k in the second sample can generate spuriously large estimates of N. If $k = 0$, which occurs several times in the trial, no estimate is possible (these instances are assigned a value of $N = 0$).

These results can be regarded as quite decent, given the minimal assumptions made, and will hopefully be better than pure guesswork.

6.5 Probability of a Nuclear Disaster

Considerable effort has gone into assessing the probability of a serious failure in a nuclear reactor, such as a significant escape of radiation or complete meltdown. In 1985, the US Nuclear Regulatory Commission gave a value of 3×10^{-4} to what we here denote μ, the probability of a serious failure, per reactor, per year [25]. More recent research has suggested that this value should be reduced to about 10^{-5} [26], although the actual value will depend on reactor type, what sort of accident is considered 'serious', and will come with some associated uncertainty.

The problem to be looked at here is, though, how one works out the probability of occurrence of at least one accident worldwide, over a number of years. As of 2021, there are 443 reactors distributed across many countries, each expected to operate for at least several decades [27], so we can expect the probability of a serious accident to be somewhat larger than is given by μ alone.

To derive the operating equation, treat first the situation of one reactor operating over, say, three years. We might think that the cumulative probability is simply 3μ, but this won't work because it does not take full account of the various ways in which an accident *could* occur over that 3-year span. This takes us into the area of combinatorics; for example, there could be an accident in the first year, but not in the second or third. Or accidents could occur in the first two years but not in the third. And so on. An implicit assumption here is that the same reactor can suffer several serious accidents in the 3 year span and still remain operational.

If the probability of at least one accident over the 3 year period is designated p_3, then this will be given by the following sum of constituents:

$$\begin{aligned}p_3 &= 3\mu\left(1-\mu\right)^2 \quad \text{(from year 1 } or \text{ year 2 } or \text{ year 3),}\\&+\mu^2\left(1-\mu\right) \quad \text{(from years 1 } and \text{ 2),}\\&+\mu^2\left(1-\mu\right) \quad \text{(from years 1 } and \text{ 3),}\\&+\mu^2\left(1-\mu\right) \quad \text{(from years 2 } and \text{ 3),}\\&+\mu^3 \quad \text{(from years 1 } and \text{ 2 } and \text{ 3),}\\&= 3\mu - 3\mu^2 + \mu^3,\end{aligned}$$

the accompanying annotation defining the relevant constituent combination. The presence of the $(1-\mu)$ factor in the first four contributions requires some explanation. What we are actually saying, in the case of an accident in *just* year 1, is that there will be *no* accidents in years 2 or 3, and since the probability of no accident is $1-\mu$, the correct constituent probability will be $\mu\left(1-\mu\right)^2$. And then multiply by 3 since 'just year 2' or 'just year 3' will have the same probability.

It can be appreciated that going through this combinatoric process for more than a few years becomes rather tedious. Although there are patterns that can be exploited to simplify the logic [14], in the present context there is a useful short-cut. This looks first for the probability that there will be *no* accidents in any of the three years — meaning no accident in year 1 and none in year 2 and none in year 3 — giving

$$p_{n3} = (1-\mu)^3,$$

the subscript $_{n3}$ denoting 'none in 3 years'. Then turn this around: the opposite of 'none in 3 years' will be 'at least one' in that duration, so

$$p_3 = 1 - (1-\mu)^3.$$

More generally,

$$p_y = 1 - (1-\mu)^y,$$

where y counts the number of years.

We are now in a position to determine the cumulative probability of accident from m reactors over y years. Using the same arguments as above, and treating

6.6. WEAPON-TARGET ALLOCATION

individual reactor accidents as independent, we get:

$$p_{mr} = 1 - (1 - p_y)^m,$$
$$= 1 - \left[1 - \left\{1 - (1-\mu)^y\right\}\right]^m,$$
$$= 1 - (1-\mu)^{my}. \qquad (6.6)$$

We can work the other way round as well — looking first at the probability of at least one accident per year for m reactors, and then the probability over y years — and come up with the same equation.

Now have a look at working out the probability of at least one accident worldwide over, say, 20 years. Then

$$p_{mr} = 1 - (1-\mu)^{443 \times 20} = 0.085,$$

assuming $\mu = 10^{-5}$. This is thankfully quite small — about 8% — but increases to a more worrying value of nearly 60% if $\mu = 10^{-4}$ instead.

The variation of p_{mr} with μ is plotted in Figure 6.5, using $m = 443$ and $y = 20$, and with a logarithmic μ-scale.

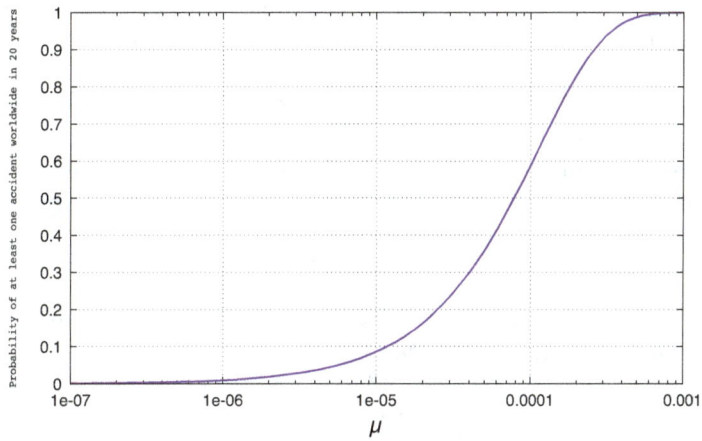

Figure 6.5: Probability of at least one accident worldwide in 20 years

Given an acceptable (preferably low) level of risk, one could use this curve, or ones like it, to determine what the design value of μ should be. Alternatively, equation (6.6) can easily be inverted to give

$$\mu = 1 - (1 - p_{mr})^{1/(my)}. \qquad (6.7)$$

6.6 Weapon-Target Allocation

This particular application of probability goes back to the late 1950s and provides a basis for optimally allocating multiple defensive weapons in an engagement with incoming attacking entities. As a specific example, numerous incoming attack aircraft of various types need to be engaged using defensive anti-aircraft missiles, again with various capabilities. It is expected that the performance of any interceptor missile will be less than perfect and can be modelled using a 'kill probability' p. In addition, each incoming attacking aircraft is assumed to have a relative importance value V assigned to it, which may (for instance) reflect its expected destructive capability. On this basis, the expected damage such an aircraft can inflict can be encapsulated in the expression

$$\text{damage} = V(1-p).$$

The $1-p$ quantity defines the probability of *not* destroying the aircraft and the ensuing damage will then be a maximum if $p=0$ and a minimum if $p=1$.

When dealing with multiple attacking aircraft and multiple defensive missiles, the damage equation takes the form of a cost function [28],

$$D = \sum_{j=1}^{n} V_j \prod_{i=1}^{m} (1-p_{ij})^{\mu_{ij}}, \tag{6.8}$$

which looks more complicated than it actually is. Here, p_{ij} is the kill probability of missile i against aircraft j, while V_j is the importance of aircraft j (its damage capability, say). It is assumed that there are n incoming aircraft, hence the summation over j, and m interceptor missiles.

The quantity $\prod$ may be unfamiliar; this is the equivalent of $\sum$ but for products rather than summations. So $\prod_{k=1}^{n} k = 1 \times 2 \times 3 \times \ldots \times n$ (thus being the same as $n!$). So if, for example, there is one incoming aircraft and we have two missiles that can be launched at it, the cost function simplifies to

$$D' = V_1 (1-p_{11})^{\mu_{11}} (1-p_{21})^{\mu_{21}},$$

which just defines the penalty (potential damage) of not hitting the aircraft with missile 1 *and* not hitting it with missile 2.

So far, the quantities described are either known or can be estimated, whereas μ_{ij} needs to be assigned as part of the optimisation process. It is set to unity if missile i is assigned to aircraft j, or zero if not so assigned[8].

The aim, then, is to choose μ_{ij} such that the cost function D is a minimum. To continue with the simpler situation D', it can be seen that there is merit in firing both missiles at the incoming aircraft (*i.e.* setting $\mu_{11} = \mu_{21} = 1$), since the product of the probability terms will always be less than their individual numeric values[9].

For small numbers of aircraft and missiles (perhaps a few tens or maybe hundreds of each, depending on the computer capabilities), exact enumerative solutions are possible in real time, thus allowing for an effective and timely defence. However, it is known that the time taken to compute such an optimal solution grows exponentially with the problem size [29], thus giving rise to the need for various sub-optimal (but hopefully near-optimal) solution techniques.

---

[8]In a more general sense, values of μ_{ij} greater than unity can be accommodated if i and j describe difference *types* of missiles and aircraft rather than a simple enumeration over each.

[9]Excluding the extremal probability values of zero or unity.

6.6. WEAPON-TARGET ALLOCATION

A different asset-based formulation of the assignment problem can be found in [30], and in this case the cost function is written

$$S = \sum_{k=1}^{K} A_k \prod_{j=1}^{n} \left[1 - q_{jk} \prod_{i=1}^{m} (1 - p_{ij})^{\mu_{ij}} \right].$$

The sum over k is over the defended assets, each of which is assigned some value A_k, and the aim is to maximise the survival of these, which means choosing the μ_{ij} so as to maximise S. The other quantity that has been introduced here is q_{jk}, which is the probability that aircraft j destroys asset k.

Chapter 7

CALCULUS

T is difficult to travel very far in mathematics without bumping into calculus in one form or another, since it forms the foundation for so much else. There is, perhaps, a tendency to regard it as too complicated to tackle, a perception that was addressed in a very readable book by Sylvanus Thompson [31] — first published in 1910 and still in print (my own copy is from 1918 and now definitely showing its age ...).

He starts by stating the common-sense meaning of such symbols as dx, which stands for 'a little bit of x', and the integral symbol $\int$, meaning 'sum of' (this symbol originated as a 'long s', as was used in antiquated books and manuscripts). This is a different sort of sum from the $\sum$ encountered in Section 4.2, which dealt with sums of discrete entities; $\int$ is summing over large (in the limit, infinite) numbers of infinitesimal quantities.

7.1 Slope of a Curve

So the expression dy/dx means, in common-sense terms, the ratio of a small bit of y divided by a small bit of x. This has immediate application to the slope of a curve, as illustrated in Figure 7.1.

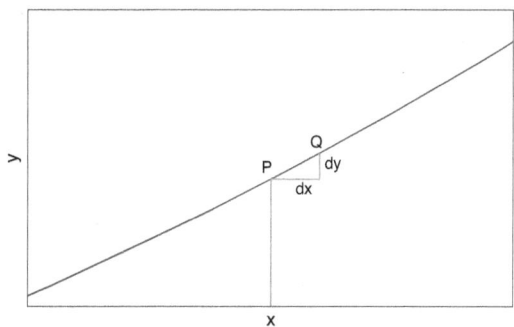

Figure 7.1: Slope of a curve

We can imagine here taking two points P and Q on the curve, a short distance dx apart in x and dy apart in y. The ratio will then give an approximation to the slope of the curve, if the curve was actually a straight line joining P and Q. The neat bit now is to imagine progressively moving Q down the curve towards P and so shrinking dx smaller and smaller towards zero; dy will shrink at the same time. In the limit as dx goes to zero, we will obtain the slope exactly at P, as desired.

By way of illustration, suppose $y = x^2$, as in Figure 7.1. Then,

$$\begin{aligned} dy &= (y \text{ at } Q) - (y \text{ at } P), \\ &= (x + dx)^2 - x^2, \\ &= 2x\,dx + dx^2. \end{aligned}$$

Dividing this by dx then gives:

$$\frac{dy}{dx} = 2x + dx,$$

and as dx tends to zero, the slope is found to be

$$\frac{dy}{dx} = 2x.$$

Referring back to Section 3, dy/dx is also equal to the tangent of the angle at P; see Figure 3.1.

In a similar manner, it is not difficult to show that

$$\frac{d}{dx}x^n = nx^{n-1},$$

for some n, and that the derivative of a constant is zero.

If Figure 7.1 expressed the path of some object as distance y versus time x, then the derivative dy/dx would supply the speed at the point x. Similarly, the second derivative d^2y/dx^2 would give the acceleration, or rate of change of speed.

An often-used short-hand for derivatives of y with respect to x is the 'dash' or 'prime' notation, so that

$$\frac{dy}{dx} \text{ is written } y'(x),$$

$$\frac{d^2y}{dx^2} \text{ is written } y''(x),$$

and so on. This notation will come in handy later on.

This is also a good place to introduce the standard notation for a function of one or more variables. So $y = f(x)$ means that y equals some function f of x; an example might be where f is the sine function. This notation is not restricted to one variable either: $y = f(x, t)$ means that y is now dependent on both x and t.

7.2. THE EXPONENTIAL FUNCTION

Differentiation and integration are inverses, or 'opposites', of one another. So if we have

$$\frac{dy}{dx} = f(x),$$

for some function f of x, then y will be given by

$$y = \int f(x)dx. \tag{7.1}$$

As mentioned above, the 'long s' symbol stands for 'sum', while the inclusion of the dx quantity indicates that we are summing numerous very small slices in x under the curve. This will bring us to determining the area under the curve between two x-limits, but first have a look at some useful functions.

7.2 The Exponential Function

The exponential function is so important in mathematics that a short section is worth inserting here. It is the only function whose derivative is the same as itself:

$$\frac{d}{dx}e^x = e^x. \tag{7.2}$$

This means that the slope of the function at any point is the same as the value of the function itself at that point. It also means that higher derivatives — d^n/dx^n — are again equal to e^x.

The exponential function can be written as an infinite series in powers of x, convergent for all x:

$$e^x = 1 + x + \frac{x^2}{2!} + \frac{x^3}{3!} + \frac{x^4}{4!} + \dots,$$

where the quantity $n!$ stands for $n(n-1)(n-2)(n-3)\dots 1$ — the product of all integers from 1 to n inclusive. This series expansion follows immediately from equation (7.2).

The series also gives the numeric value of e, by setting $x = 1$:

$$e = 1 + 1 + \frac{1}{2!} + \frac{1}{3!} + \frac{1}{4!} + \dots,$$
$$= 2.7182818\dots$$

and it may be inferred that $e^0 = 1$.

Let us have a look at the shape of the exponential function e^x, and also e^{-x} (which is the same as $1/e^x$). Graphs of both of them are provided in Figure 7.2; e^x can also be written $\exp(x)$.

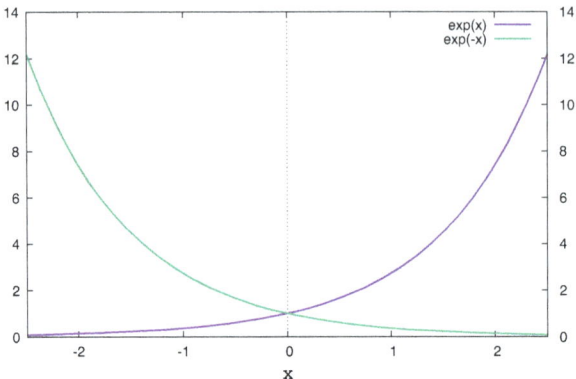

Figure 7.2: Exponential functions

For e^x, the function values go to zero for large negative x and to infinity (rapidly) as x increases in the positive direction. The functions are mirror images of one another.

To return to the derivative of the exponential function, a useful result that will be needed later is:
$$\frac{d}{dx}e^{\lambda x} = \lambda e^{\lambda x},$$
if λ is independent of x.

7.3 Trigonometric Functions

We have already met the trigonometric functions in Chapter 3, but a brief diversion is useful here to explain where the derivatives come from and to add in their series expansions. Work first from the known equations (3.5) and (3.6) regarding sums of angles:
$$\sin(\theta + \phi) = \sin\theta\cos\phi + \cos\theta\sin\phi,$$
$$\cos(\theta + \phi) = \cos\theta\cos\phi - \sin\theta\sin\phi.$$
Set $\phi = d\theta$ for a small bit of θ. So we can follow essentially the same approach as in the case of x^2 above, making use of the fact that $\cos d\theta \approx 1$, $\sin d\theta \approx d\theta$ for small $d\theta$ (these approximations follow by imagining Figure 3.1 to be an extremely skinny triangle). The end results are:
$$\frac{d}{d\theta}\sin\theta = \cos\theta \text{ and } \frac{d}{d\theta}\cos\theta = -\sin\theta. \tag{7.3}$$
Provided the angle θ is measured in radians (with 2π radians in a complete circle), the infinite series functions are:
$$\sin\theta = \theta - \frac{\theta^3}{3!} + \frac{\theta^5}{5!} - \frac{\theta^7}{7!} + \ldots,$$
$$\cos\theta = 1 - \frac{\theta^2}{2!} + \frac{\theta^4}{4!} - \frac{\theta^6}{6!} + \ldots.$$

7.4. HYPERBOLIC SINE AND COSINE FUNCTIONS

Differentiating each of these series term by term and cross-comparing, again results in equations (7.3), as expected.

Graphs of the sine and cosine functions are given in Figure 7.3.

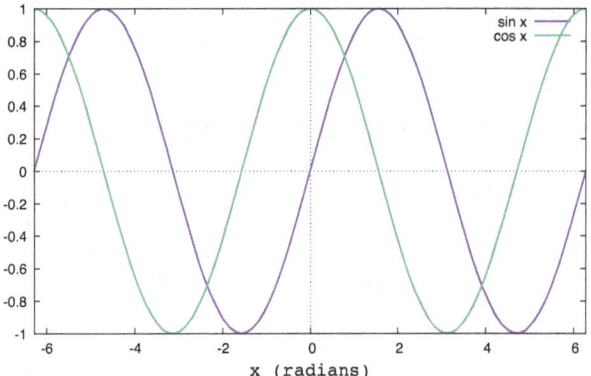

Figure 7.3: Trigonometric functions

With a bit of visual imagination, it is evident that either curve can be shifted to the left or right to overlay the other one. For example,

$$\cos\left(x - \frac{\pi}{2}\right) = \sin x.$$

Both functions are also periodic, coming back to the same point after shifting x by $\pm 2\pi$.

7.4 Hyperbolic Sine and Cosine Functions

These are defined as follows:

$$\cosh x = \frac{1}{2}\left(e^x + e^{-x}\right),$$
$$\sinh x = \frac{1}{2}\left(e^x - e^{-x}\right),$$

the 'h' in the names denoting 'hyperbolic'. It might be expected that these functions are related in some way to the more familiar trigonometric functions (and this is the case), but at present their derivatives are of most interest. That is,

$$\frac{d}{dx}\cosh x = \sinh x,$$
$$\frac{d}{dx}\sinh x = \cosh x,$$

which are easily obtained from the function definitions.

It may be remembered from Section 3 that $\cos^2 \psi + \sin^2 \psi = 1$. Well, there is a similar relation joining $\cosh x$ and $\sinh x$, which states that

$$\cosh^2 x - \sinh^2 x = 1,$$

and this also follows from the function definitions.

7.5 Taylor and Maclaurin Series

The polynomial expansions that we met in Sections 7.2 and 7.3 prompts the question: can *any* function be written in such a form? The answer is yes, subject to some constraints, so proceed as follows: suppose we have some function $f(x)$ that has at least n derivatives at the point $x = 0$. Then try a polynomial of degree n:

$$f(x) = b_0 + b_1 x + b_2 x^2 + b_3 x^3 + \ldots + b_n x^n,$$

for unknown constant coefficients $b_0, \ldots, b_n$.

Set $x = 0$ on both sides of the above equation, giving $b_0 = f(0)$.

Differentiate both sides and set $x = 0$, giving $b_1 = f'(0)$.

Twice-differentiate both sides and set $x = 0$, giving $2b_2 = f''(0)$, or $b_2 = f''(0)/2$.

Continuing this process results in the following expansion:

$$f(x) = f(0) + f'(0)x + f''(0)\frac{x^2}{2!} + f'''(0)\frac{x^3}{3!} + \ldots + f^{(n)}(0)\frac{x^n}{n!}.$$

This is known as the Maclaurin[1] series.

It is also possible to evaluate the derivatives at some point a instead of at zero, resulting in the Taylor[2] series [9]:

$$f(x) = f(a) + f'(a)(x-a) + f''(a)\frac{(x-a)^2}{2!} + f'''(a)\frac{(x-a)^3}{3!} + \ldots$$
$$\ldots + f^{(n)}(a)\frac{(x-a)^n}{n!}.$$

Now define $x - a = h$, so that $x = a + h$. Then, an equivalent form of the Taylor series is:

$$f(a+h) = f(a) + hf'(a) + \frac{h^2}{2!}f''(a) + \frac{h^3}{3!}f'''(a) + \ldots + \frac{h^n}{n!}f^{(n)}(a), \quad (7.4)$$

which is the form I find most general and the easiest to remember. Such expansions form the foundation for further calculus-based results (such as the radius of curvature in Section 7.9).

---

[1] Colin Maclaurin, 1698 to 1746.
[2] Brook Taylor, 1685 to 1731.

7.6 Area Under the Curve

Consider the green shaded area shown in Figure 7.4.

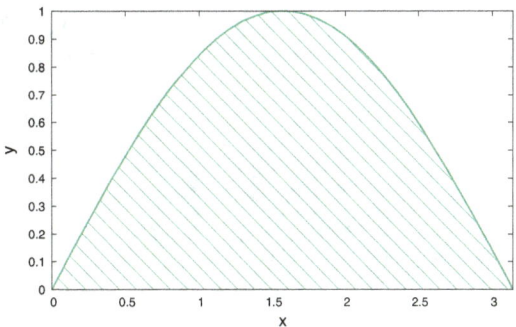

Figure 7.4: Example area

This represents the area under a half-cycle sine curve, from zero to π in radians (equivalent to zero to 180°).

Using the above integral results, the area under the curve in Figure 7.4 is given by:

$$A = \int_0^\pi \sin x \, dx. \tag{7.5}$$

This is termed a *definite integral*, and differs from equation (7.1) in that we now have lower and upper bounds on x — specifically, zero and π.

Recall from the preceding example that the derivative of x^2 is $2x$, and that differentiation and integration are inverses of each other. So, to work out A, we need a function that, when differentiated, will give $\sin x$. This is known to be $(-\cos x)$ (noting the negative sign), so

$$A = \left[-\cos x\right]_0^\pi = -\cos \pi + \cos 0 = 2,$$

since $\cos \pi = -1$ and $\cos 0 = 1$.

It may be remarked that although differentiating a function is generally easy, integration can be much harder. In some cases, numerical integration is necessary, and the simplest approach is to subdivide the required length of integration into numerous small subsections of equal length h. Continuing with the above example, suppose that there are $n = \pi/h$ subsections; then replace the integration with a summation[3]:

$$\int_0^\pi \sin x \, dx \longrightarrow h \sum_{j=0}^{n-1} \sin x_j, \text{ with } x_j = jh.$$

---

[3] The upper limit is $n-1$ since the y-value for each sub-interval is represented by its left-hand boundary.

This not especially accurate, and care is needed in the choice of h for rapidly-varying functions, but as a 'quick hack' it is a useful approach. More accurate methods for numerical integration can be found in [32].

Integration can also be used to determine volumes, using essentially the same 'divide and conquer' approach. For example, suppose we have a cauldron-like shape such as that shown in Figure 7.5, in which there is axial symmetry.

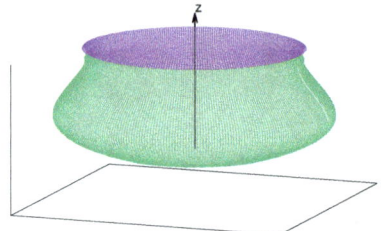

Figure 7.5: Volume of revolution, perspective view

Since the cross-section in the axial direction is known to be circular, the solution is given by the integral:

$$V = \int_{z=0}^{Z} \pi \left\{ f(r) \right\}^2 dz,$$

which has divided the volume up into a large number of thin discs, each disk with radius $f(r)$ and with thickness dz (denoted by the black line in Figure 7.6).

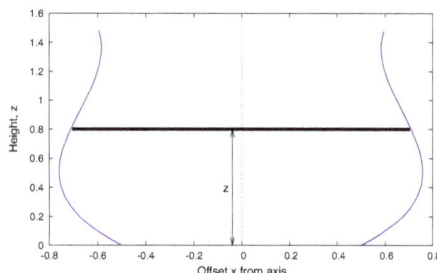

Figure 7.6: Volume of revolution, cross-section

Here, r is the magnitude of the offset x from the longitudinal axis and Z is the height of the vessel. This does assume that the curved shape $z = f(r)$ of the vessel wall is known or can be approximated.

7.7 Maxima and Minima

An important corollary to determining the slope of a curve is the ability to locate the maxima and minima — the extremal points — of a function. Take a look at Figure 7.7.

7.8. PATH LENGTH ALONG A CURVE

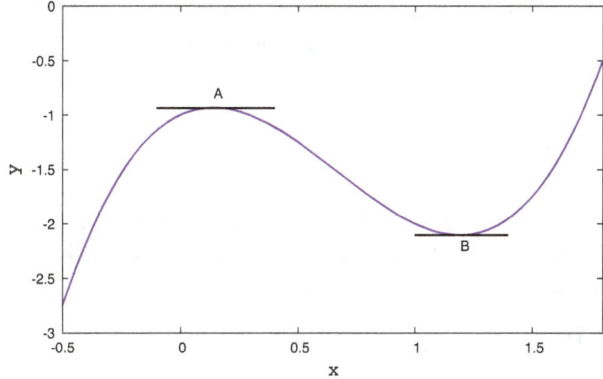

Figure 7.7: Maximum and minimum of a cubic curve

This is a graph of the cubic function $y = 2x^3 - 4x^2 + x - 1$. By inspection it can be seen that the function has a maximum at point A and a minimum[4] at point B. In both cases, the slope of the curve is zero, so the extremal points are given by $dy/dx = 0$.

Taking the second derivative of the function, d^2y/dx^2, will determine whether an extremal point is a maximum or a minimum: if negative, it is a maximum; if positive, it is a minimum. If the second derivative is zero, we have a *point of inflection*, which corresponds to neither a maximum nor a minimum but to something in between.

The ability to locate the maximum or minimum of some quantity has broad application, as will be seen in later sections.

7.8 Path Length Along a Curve

The path length along a curve invokes Pythagoras' theorem again. Look back at Figure 7.1 and write

$$ds^2 = dx^2 + dy^2,$$

for the infinitesimal right-angled triangle. The parameter s measures length along the curve, and so

$$s = \int_a^b ds = \int_a^b \sqrt{dx^2 + dy^2},$$

between some limits a and b. This can also be written as:

$$s = \int_{x_1}^{x_2} \sqrt{1 + \left(\frac{dy}{dx}\right)^2}\, dx,$$

---

[4]Larger and smaller values of y can of course be found, but since these are unbounded in magnitude they do not correspond to definite extremal points.

with both the integration and limits in terms of x.

So in the case of Figure 7.4, the length of the upper arc boundary from zero to π will be given by:

$$s = \int_0^\pi \sqrt{1 + \cos^2 x} \, dx,$$

which has a solution in terms of elliptic integrals but otherwise needs to be evaluated numerically. An approximate value to two decimal places is 3.82.

7.9 Radius of Curvature

For a two-dimensional curve given by $y = f(x)$, it is possible to define the *radius of curvature* ρ at any point by the equation

$$\rho = \frac{[1 + f'^2]^{3/2}}{f''}, \qquad (7.6)$$

using the 'dash' notation from Section 7.5.

So where does this come from? It is not proposed here to go through the detailed derivation [33], which involves a fair amount of algebra. The idea is to consider three points P, Q and R on the curve that are close together (but not coincident) and then use them to define the arc of a circle such that

$$(x - a)^2 + (y - b)^2 = r^2,$$

where the centre of the circle is defined by the point (a, b) in the x-y plane and r is the circle radius[5]. Next define the three points P, Q and R as follows:

P : $(x, f(x))$,
Q : $(x + h, f(x + h))$,
R : $(x + k, f(x + k))$,

where the quantities h and k are small (and unequal). So Q and R are considered small shifts away from P. Then use the Taylor series (Section 7.5) to approximate the y coordinate of point Q:

$$f(x + h) \approx f(x) + h f'(x) + \frac{h^2}{2!} f''(x),$$

(no need to go any further with the expansion). Do the same for the y coordinate of point R, to get another similar expansion in terms of k. The coordinates for each of the three points can then be plugged in to the above equation for the circle, which enables the derivation of a, b and — most importantly — r. In the limit as h and $k \to 0$, r becomes ρ, the quantity of interest.

---

[5]That this *is* the equation of a circle can be established by shifting to polar coordinates using $x = a + r \cos \theta$, $y = b + r \sin \theta$ and recalling that $\cos^2 \theta + \sin^2 \theta = 1$.

Returning to equation (7.6), the *curvature* κ (Greek 'kappa') is defined as the inverse of the radius of curvature, so

$$\kappa = \frac{1}{\rho} = \frac{f''}{[1 + f'^2]^{3/2}}.$$

If the magnitude of f' is known to be small, then the approximation $\kappa \approx f''$ may be used; this appears in the bending of beams, for instance (Section 12).

7.10 Additional Expressions

It is useful here to add a short diversion into various differentiation-related expressions and results. Thus, for future reference, a variation on the df/dx symbology is the following, using a curly d symbol:

$$\frac{\partial}{\partial x} f(x, t).$$

This is an example of *partial differentiation*, in which we are differentiating a function f of two independent variables, x and t. The differentiation here is with respect to x with t held constant[6].

A related result that will be needed later is when differentiating something like $f(x(t), y(t))$, in which function f depends on both x and y, both of which in turn depend on t. Then

$$\frac{df}{dt} = \frac{\partial f}{\partial x}\frac{dx}{dt} + \frac{\partial f}{\partial y}\frac{dy}{dt}. \tag{7.7}$$

Consider also the differentiation of a *product* of two functions: $f(t) = u(t)v(t)$. Then,

$$\frac{df(t)}{dt} = u(t)\frac{dv(t)}{dt} + \frac{du(t)}{dt}v(t).$$

This has application to integration; write it in the form:

$$d\left(u(t)v(t)\right) = u(t)\frac{dv(t)}{dt}dt + \frac{du(t)}{dt}v(t)dt,$$

and integrate both sides with respect to t, thus;

$$\int d\left(u(t)v(t)\right) dt = \int u(t)\frac{dv(t)}{dt}dt + \int \frac{du(t)}{dt}v(t)dt.$$

This implies that

$$u(t)v(t) = \int u(t)\frac{dv(t)}{dt}dt + \int \frac{du(t)}{dt}v(t)dt,$$

---

[6] Some other partial-differential-related symbologies that the reader might encounter are u_x, standing for $\partial u/\partial x$ (popular in texts on fluid dynamics), and ∂_μ which is short-hand for $\partial/\partial x_\mu$ (this one occurs in quantum physics and related areas).

or, rearranging,

$$\int u(t)\frac{dv(t)}{dt}dt = u(t)v(t) - \int \frac{du(t)}{dt}v(t)dt. \tag{7.8}$$

This is a valuable way of evaluating some integrals.

And finally here, if $g(t) = u(t)/v(t)$, then

$$\frac{dg}{dt} = \frac{1}{v(t)}\frac{du(t)}{dt} - \frac{u(t)}{v^2(t)}\frac{dv(t)}{dt}. \tag{7.9}$$

Chapter 8

BICYCLE WHEEL REFLECTORS

WHEN our daughter was young, her bicycle wheels had little plastic reflectors attached to the spokes, which made a pretty pattern when she cycled along in the sunshine. As a somewhat whimsical example of the application of calculus, we ask what sort of curve in space does each such reflector make?

To create a well-defined example, assume that the reflector is as near to the circumference of the wheel as makes no difference, and that the bicycle is moving from left to right at a constant speed v. The situation is illustrated in Figure 8.1, with the green circle denoting the bicycle wheel at the start, and the purple line marking the spatial path of the reflector on the rim. The purple line is called a *cycloid*.

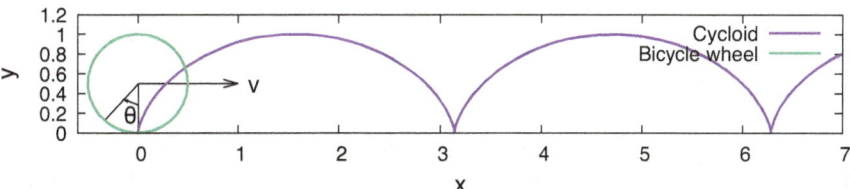

Figure 8.1: Cycloid curve

The following derivation is a bit roundabout in nature, but is intended to be easy to follow. Assume that at the start the reflector is on the ground at the point $(0,0)$ in Figure 8.1 and that the angle θ is measured clockwise from this point (since the wheel is moving to the right). The wheel itself is assumed to have a diameter of 1 metre.

Temporarily imagine the wheel to have lost contact with the ground and be spinning freely. Then, working from the illustrated angle θ, the x and y coordinates of the reflector will be given by:

$$x = -a \sin \theta,$$
$$y = a - a \cos \theta.$$

These follow from the right-angled triangle rules defined in Chapter 3, although bear in mind that x is measured positive to the right and y from the bottom of the wheel rather than its centre. Here, a is the wheel radius (0.5 metres).

Differentiate these with respect to time, to get:

$$\frac{dx}{dt} = -a\frac{d\theta}{dt}\cos\theta,$$
$$\frac{dy}{dt} = a\frac{d\theta}{dt}\sin\theta,$$

since a is a constant[1] but θ increases with time as the wheel rotates.

These equations for the velocity are for a freely spinning wheel, with no ground contact. Now allow the wheel as a whole to move to the right with a speed v, but still with no ground contact. The velocity components then become:

$$\frac{dx}{dt} = -a\frac{d\theta}{dt}\cos\theta + v,$$
$$\frac{dy}{dt} = a\frac{d\theta}{dt}\sin\theta,$$

noting that there is no vertical motion for the wheel as a whole (a flat ground is assumed).

Up to this point, the speed v and the rotation rate $d\theta/dt$ are completely decoupled and can be defined independently. To link the two, as is the case where the wheel makes ground contact with no slipping, it is necessary to set the speed of the reflector to zero when $\theta = 0$. Since the ground is not moving, and the wheel is not slipping, contact between the two must mean that the lateral speed of the contact point is instantaneously zero. Thus we get, at $\theta = 0$,

$$-a\frac{d\theta}{dt} + v = 0, \text{ which implies that } v = a\frac{d\theta}{dt}. \tag{8.1}$$

In effect, the no-slip condition means that the wheel rotation rate is driven entirely by the speed of the bicycle as a whole. Intuitively obvious, perhaps (especially to a cyclist), but still worth going through in a bit more detail than is strictly necessary.

So we now have for the velocity components:

$$\frac{dx}{dt} = -a\frac{d\theta}{dt}\cos\theta + a\frac{d\theta}{dt}$$
$$\frac{dy}{dt} = a\frac{d\theta}{dt}\sin\theta.$$

Re-integrate these, at the same time imposing the initial conditions, to obtain:

$$x = a(\theta - \sin\theta),$$
$$y = a(1 - \cos\theta).$$

---

[1] Hopefully the wheel is actually round.

Note that the time coordinate, which was introduced in the manner of a catalyst to bring in the bicycle speed v, has disappeared. These two equations give the spatial coordinates x, y of the reflector in *parametric form*, θ being the linking parameter. Increasing θ over the range from zero radians to 20 radians then gives rise to the purple line in Figure 8.1.

The above analysis is for a point actually on the rim of the wheel, where it will be in contact with the ground. But what about the more usual situation, where the reflector is set on one of the spokes, some way in from the circumference?

Assume the reflector distance from the wheel centre is now b, with $b < a$. Following through the above arguments, but with the initial x and y given in terms of b rather than a, and remembering that the link between v and $d\theta/dt$ is still given by equation (8.1), we find that

$$x = a\theta - b\sin\theta,$$
$$y = a - b\cos\theta.$$

For a reflector located 3/4 of the way out from the wheel centre, its path in space will appear to an external observer as in Figure 8.2.

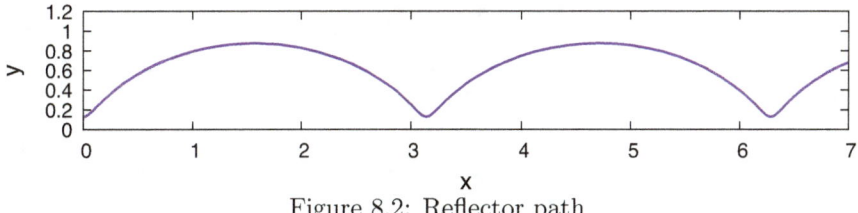

Figure 8.2: Reflector path

With this location of the reflector, it does not come to an instantaneous stop at $\theta = 0$, so the cusps in the curve that were evident in Figure 8.1 are now absent.

Chapter 9

LADDER IN A MINE SHAFT

ANOTHER simple example of what calculus can do is borrowed from [34], and is concerned with the maximum length of a ladder that can be manœuvred around a sharp bend in a mine tunnel. The situation is illustrated in Figure 9.1.

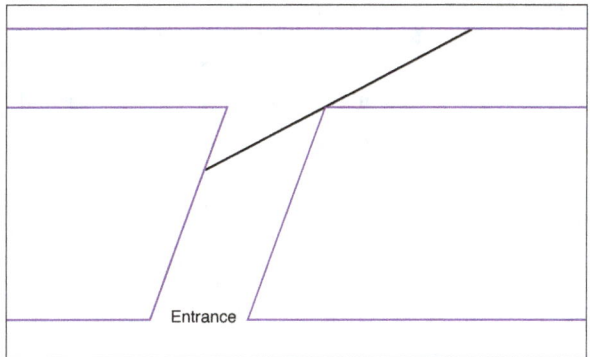

Figure 9.1: Manœuvring a ladder into a mine-shaft

The idea is to move the ladder from the entrance shaft into the long gallery at the top of the picture, and the obvious first obstacle is the angle at which the tunnels join. One can imagine trying out various positions, with the main sticking point as illustrated: the lower end of the ladder jammed up against the left-hand wall of the entrance, the further end of the ladder touching the further cross-tunnel wall, and some ladder mid-point constrained by the angle in the wall.

It can be assumed that pruning bits off the ladder until the junction can be successfully negotiated is not an option, so the intent is to find out in advance what length of ladder can be accommodated given the tunnel geometry. Adding in some notional lengths and angles gives Figure 9.2.

59

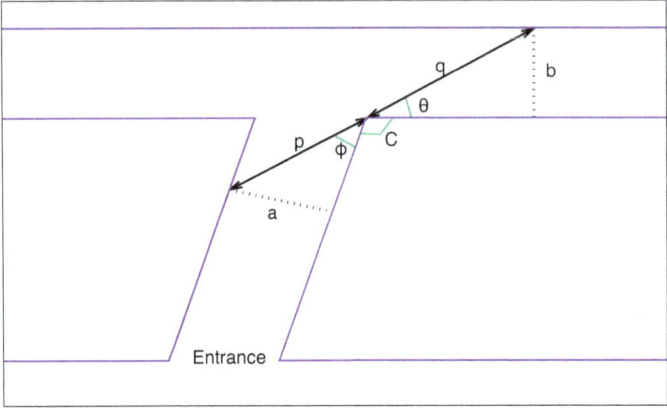

Figure 9.2: Geometry of the ladder in the mine-shaft

The tunnel widths are marked as a for the entrance and b for the cross-shaft[1], with the junction angle between the two shafts defined as C. The ladder has also been visually split into two, with section lengths p and q. So the total length of the ladder is $L = p + q$. The other two angles θ and ϕ depend on how the ladder is positioned relative to the corner, and $\phi + C + \theta = \pi$ radians (equivalent to $180°$). The ladder thickness is considered negligible relative to the tunnel dimensions.

We can now use the right-angled triangle trigonometric definitions from Chapter 3 to obtain the following:

$$\sin\phi = \frac{a}{p}, \quad \sin\theta = \frac{b}{q},$$

which can be rearranged to give

$$p = \frac{a}{\sin\phi} \quad \text{and} \quad q = \frac{b}{\sin\theta}.$$

Therefore,

$$L = \frac{a}{\sin\phi} + \frac{b}{\sin\theta},$$
$$= \frac{a}{\sin(\pi - C - \theta)} + \frac{b}{\sin\theta}. \tag{9.1}$$

This gives L as a function of θ, plus assorted tunnel-dependent quantities. From Section 7.7, a maximum or minimum of L will then be given by setting the differential with respect to θ to zero:

$$\frac{dL}{d\theta} = 0.$$

---

[1]In whatever units are relevant.

In order to differentiate something like $b/\sin\theta$, we need equation (7.9), with u constant, so that:

$$\frac{d}{d\theta}\left(\frac{b}{\sin\theta}\right) = -\frac{b\cos\theta}{\sin^2\theta}.$$

In a similar manner,

$$\frac{d}{d\theta}\left(\frac{a}{\sin(\pi-C-\theta)}\right) = \frac{a\cos(\pi-C-\theta)}{\sin^2(\pi-C-\theta)}.$$

Putting the bits together then gives:

$$\frac{dL}{d\theta} = \frac{a\cos(\pi-C-\theta)}{\sin^2(\pi-C-\theta)} - \frac{b\cos\theta}{\sin^2\theta},$$
$$= 0 \text{ for an extremal.} \qquad (9.2)$$

This is known as a *transcendental equation* and needs to be solved numerically. But before delving into a candidate numerical algorithm, graphing the function for various values of θ not only provides insight into the behaviour of the function, but will also give the zeros (at least approximately, depending on the detail of the graph). We assume $a = 3$, $b = 2$ metres and $C = 2.02458$ radians (116°). Plotting both L and $dL/d\theta$ for various values of θ then gives Figure 9.3.

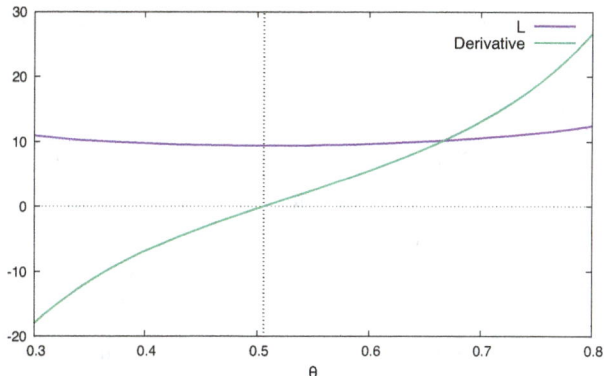

Figure 9.3: Graph of L and $dL/d\theta$ versus θ

The vertical dotted line marks where $dL/d\theta = 0$, and this corresponds to $\theta = 0.505669$ radians (or 29.9727°) and $L = 9.355$ metres for the maximum ladder length. The span of θ is restricted in Figure 9.3 since both L and its derivative are unbounded if $\theta = 0$ or if $\theta = \pi - C$.

Does this maximum value of L make sense, since it appears optimistically long given the tunnel widths? One way of checking is to generate a whole series of candidate ladder positions, each with L set to the above maximum value and show pictorially that the solution is possible. The same picture also shows how the manœuvre can take place — see Figure 9.4.

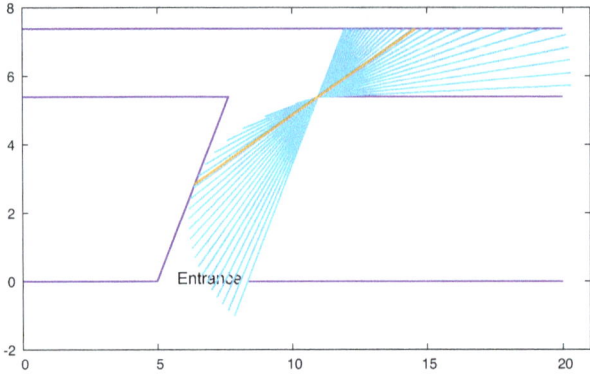

Figure 9.4: Manœuvring a ladder of $L = 9.355$ m length

The idea is to start with the ladder lying against the right-hand side of the entrance tunnel and as far in as it will go. Then use the right-hand junction corner as a pivot and rotate the ladder progressively clockwise, at the same time moving it inward. The critical point is the alignment coloured brown, when the left-hand of the ladder just brushes the left wall of the entrance tunnel. Even a slightly longer ladder will be too long and will jam against the wall[2].

The graphical solution for finding the maximum L is all very well, but for better solutions we need to be able to solve equation (9.2) to arbitrary accuracy. One commonly-used iterative technique, out of the many methods available, is due to Newton[3] and goes as follows:

1. Write the function to be solved for θ as $f(\theta) = 0$.

2. Assume that an approximate solution is given by $\theta = \theta_0$.

3. Suppose then that the *correct* solution will be obtained by adjusting θ_0 slightly, so that $\theta_1 = \theta_0 + \delta\theta$ for small $\delta\theta$. That is, $f(\theta_0 + \delta\theta) = 0$. [The unitary quantity $\delta\theta$ has a similar meaning to $d\theta$, and $\Delta\theta$ may also be used.]

4. Expand this in a Taylor series (Section 7.5), so that

$$f(\theta_0 + \delta\theta) \approx f(\theta_0) + \delta\theta f'(\theta_0) + O(\delta\theta^2) = 0,$$

where the f' denotes the first derivative of f with respect to θ, evaluated at θ_0 and ignore the higher-order terms. (The notation $O(.)$ stands for 'order of magnitude' [35]).

5. Therefore,

$$\delta\theta \approx -\frac{f(\theta_0)}{f'(\theta_0)},$$

---

[2]You can convince yourself that this is so by making a scale diagram on a sheet of paper, and trying the manœuvres out using a thin straw or similar.

[3]Isaac Newton, 1643 to 1727.

thus enabling a correction to be made to θ_0 (the curly equals sign stands for 'approximately equal to'). This new approximation can be further refined by going back to step 3 and the process then repeated until $\delta\theta$ is sufficiently small.

The method does need to be used with some care — if the initial estimate is too far off, the algorithm may diverge (or converge to another zero of the function, if such is present).

In the current example, we have

$$f(\theta) = \frac{a\cos(\pi - C - \theta)}{\sin^2(\pi - C - \theta)} - \frac{b\cos\theta}{\sin^2\theta},$$

$$f'(\theta) = \frac{a}{\sin(\pi - C - \theta)}\left\{1 + \frac{2\cos^2(\pi - C - \theta)}{\sin^2(\pi - C - \theta)}\right\} + \frac{b}{\sin\theta}\left\{1 + \frac{2\cos^2\theta}{\sin^2\theta}\right\}.$$

Starting the iterations with $\theta_0 = 0.5$ then generates the sequence of values listed in Table 9.1.

Table 9.1: Iterative solutions for θ

Iteration	θ	$\delta\theta$
0	0.50000000	0.00565166
1	0.50565166	1.76508×10^{-5}
2	0.50566932	1.48553×10^{-10}

Chapter 10

LENGTH OF A SPIRAL

I had occasion recently to apply draught excluder to one of the back windows, where the east wind had sought out an ill-fitting section. You probably know the kind of stuff I mean: a thin strip of foam with a sticky backing, the whole lot wound into a coil around a cardboard ring. After viewing the much-depleted coil, it belatedly occurred to me to ask how the manufacturer managed to work out the length of foam corresponding to a given outer radius or package size.

A suitable mathematical model for such a coil is an Archimedean[1] spiral, for which the radius at any point is simply proportional to the angle. That is,

$$r = r_0 + \kappa\theta, \tag{10.1}$$

in which κ is a constant. See Figure 10.1; the purple circle of radius r_0 at the centre corresponds to the cardboard ring.

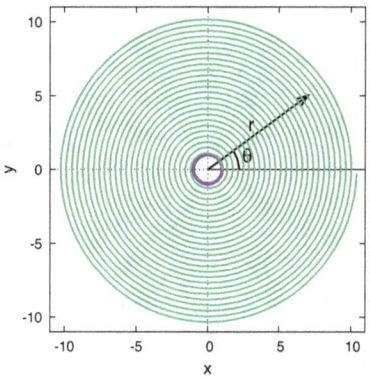

Figure 10.1: Archimedean spiral

This simple equation has the property that the radius increases by a fixed amount on every θ-cycle, since $dr/d\theta$ is a constant. The parameter κ relates to

---
[1] After Archimedes, c. 287 to 212 BC.

the thickness τ of each layer in the coil via the equation $\tau = 2\pi\kappa$; this can be appreciated by looking at the change in r along any radial line as θ increases by another cycle of 2π.

In terms of Cartesian coordinates, the path length along any infinitesimal section of the spiral will be given by ds, where

$$ds^2 = dx^2 + dy^2 \text{ (see Section 7.8).}$$

In polar coordinates, which are easier to work with in the present instance, this becomes

$$ds^2 = dr^2 + r^2 d\theta^2,$$

which may be derived using $x = r\cos\theta$, $y = r\sin\theta$. So the required length L of the spiral-wound insulation is given by

$$L = \int_{r=r_0}^{r_1} ds = \int_{r=r_0}^{r_1} \sqrt{dr^2 + r^2 d\theta^2},$$

where r_1 is the maximum radius of the coil.

Given equation (10.1), it is convenient to work entirely in terms of r using $d\theta = dr/\kappa$, so that

$$L = \int_{r=r_0}^{r_1} \sqrt{1 + r^2/\kappa^2}\, dr,$$

and to simplify the following analysis define $\xi = r/\kappa$, with $\xi_0 = r_0/\kappa$, $\xi_1 = r_1/\kappa$ so that

$$L = \kappa \int_{\xi_0}^{\xi_1} \sqrt{1 + \xi^2}\, d\xi = \kappa \Big[J(\xi_1) - J(\xi_0) \Big],$$

here defining the function $J(\xi)$ as the indefinite[2] integral

$$J(\xi) = \int \sqrt{1 + \xi^2}\, d\xi. \tag{10.2}$$

Faced with this expression, we can either look for the solution in a comprehensive table of integrals such as [36] (assuming a solution exists), or else have a go from first principles. The latter option will be adopted here and the first step is to try and remove the square root quantity by means of a change of integration variable. There are at least two options for this; for instance we know that $\sec^2\psi = 1+\tan^2\psi$ (obtainable using the definition of $\sec\psi = 1/\cos\psi$, combined with $\cos^2\psi + \sin^2\psi = 1$ from Section 3; we'll come back to this option), but an easier route to the answer involves the hyperbolic sine and cosine functions,

---

[2] Meaning without explicit integration limits.

which are defined in Section 7.4. In particular, we know that $\cosh^2 x - \sinh^2 x = 1$, which suggests substituting $\sinh \mu$ for ξ. Therefore, since $d\xi/d\mu = \cosh \mu$, we find that

$$J(\mu(\xi)) = \int \cosh^2 \mu \, d\mu.$$

This does not appear to have gained us very much, but another identity will assist in going to the next step. That is, use the identity

$$\cosh 2\mu = 2\cosh^2 \mu - 1.$$

To see that this equation must hold, just expand out the right-hand side:

$$2\cosh^2 \mu - 1 = 2 \left[\frac{1}{4} \left(e^\mu + e^{-\mu} \right)^2 \right] - 1,$$

$$= \frac{1}{2} \left(e^{2\mu} + 2 + e^{-2\mu} \right) - 1,$$

$$= \frac{1}{2} \left(e^{2\mu} + e^{-2\mu} \right),$$

which is $\cosh 2\mu$ by definition.

Therefore,

$$J = \frac{1}{2} \int (1 + \cosh 2\mu) \, d\mu = \frac{1}{2} \left[\mu + \frac{1}{2} \sinh 2\mu \right],$$

$$= \frac{1}{2} \left[\mu + \sinh \mu \cosh \mu \right],$$

since $d/d\mu(\sinh 2\mu) = 2\cosh 2\mu$, and $\sinh 2\mu = 2\sinh \mu \cosh \mu$ from the product of their definitions in terms of the exponential functions.

It remains to shift this lot back into functions involving ξ, which results in

$$J(\xi) = \frac{1}{2} \left[\sinh^{-1} \xi + \xi \sqrt{1 + \xi^2} \right],$$

where the first term in the square brackets is the inverse hyperbolic sine function and the square root in the second term comes from expressing $\cosh \mu$ in terms of $\sinh \mu$ via $\cosh^2 \mu = 1 + \sinh^2 \mu$.

The $\sinh^{-1}$ function can be cast back into something a bit more familiar by means of its definition; that is, if $z = \sinh^{-1} \xi$, we must have

$$\xi = \sinh z = \frac{1}{2} \left(e^z - e^{-z} \right).$$

Multiply this throughout by $2e^z$ to get

$$2\xi e^z = e^{2z} - 1,$$

and rearrange into the quadratic equation

$$e^{2z} - 2\xi e^z - 1 = 0.$$

This has two possible solutions for e^z:

$$e^z = \xi \pm \sqrt{1+\xi^2};$$

(see Appendix E).

The solution with the negative sign can be dropped, since $z \to +\infty$ must give rise to $\xi \to +\infty$ from $\xi = \sinh z$, so we end up with

$$z = \log_e \left\{ \xi + \sqrt{1+\xi^2} \right\},$$

and so

$$J(\xi) = \frac{1}{2}\left[\log_e \left\{ \xi + \sqrt{1+\xi^2} \right\} + \xi\sqrt{1+\xi^2}\right], \qquad (10.3)$$

with the required spiral length being supplied by

$$L = \kappa\left[J(\xi_1) - J(\xi_0)\right], \text{ plus } \xi = r/\kappa,\ \xi_1 = r_1/\kappa,\ \xi_0 = r_0/\kappa.$$

It's now possible to plug some numbers into this to see what the original length of my coil of insulation might have been. Using a ruler, I found that $10\,\tau \approx 3.6$ cm, implying that $\kappa \approx 0.057$ cm per radian. The numbers are approximate, since the foam is not rigid and will have been compressed somewhat over time. Also, although I can measure $r_0 = 3.8$ cm, I have to estimate $r_1 \approx 14$ cm from the overall size of the packaging. Using these numbers plus a bit of calculation, L works out at about 16 m; this is not too far off the claimed original length of 15 m that is printed on the packet (I've probably slightly over-estimated r_1).

The same logic can be applied to other items that are manufactured in spiral form, such as toilet rolls or paper kitchen towels, although in such cases κ is likely to be variable along the spiral, depending on paper thickness, tightness of winding and so forth.

As a final (brief) comment in this section, suppose we go back to equation (10.2) and ask what would have happened if the substitution $\xi = \tan\psi$ had been used instead of $\xi = \sinh\mu$. Then, since $d\xi/d\psi = \sec^2\psi$, we would get

$$J(\xi(\psi)) = \int \sec^3\psi\,d\psi;$$

this can then be integrated by parts (see Section 7.10). In view of equation (10.3), though, this integral provides a link between the trigonometric functions (or inverses thereof) and logarithms, square roots, *etc.* Such equivalences are not necessarily what one would expect, but are often very useful.

Chapter 11

NEWTONIAN DYNAMICS

NEWTON'S laws of motion may be summarised as follows [18]:

1. A particle that is at rest (or in motion) will remain at rest (or in motion) until acted on by an external force.

2. The rate of change of linear momentum of a particle is equal to the total applied force.

3. Every action has an equal and opposite reaction.

With calculus and a knowledge of Newton's second law, *Force = mass × acceleration*[1], it is possible to predict the behaviours of a wide variety of objects, from projectiles, to clock pendulums, to planetary orbits and beyond. When applied to solid objects, the rotations of objects can be predicted, and consequences of Newton's laws include Navier's equations for stress-strain in solids (used in engineering and construction) and the Navier-Stokes equations for fluid flow (employed in aerodynamics and hydrodynamics).

A different angle on dynamics is discussed in Chapter 24; this *analytical dynamics* is an approach that links up with the calculus of variations (Chapter 23) and with General Relativity (Chapter 29).

11.1 Flight of a Cannonball

My father owned a small brass cannon, a similar design to those used in Nelson's navy but about four inches long. With a half-teaspoon or so of black powder[2] and a short section of wooden dowel to act as shot, this cannon would go off with a most satisfying bang, hurling the shot across the workshop amid a decent cloud

---

[1] Correctly, *force = rate of change of momentum = rate of change of mass × velocity*, but if mass is constant, we end up with *mass × acceleration*.

[2] Gunpowder.

of noxious smoke. Favoured guests were treated to a one-gun salute, although those of a nervous disposition were not always appreciative.

Somehow, it never occurred to me at the time to try and work out the muzzle velocity, but the principles are straightforward enough. Decompose the motion into coordinates x along the ground and y vertically upward (in height). The acceleration bit of Newton's equation is given by the second derivative (see above), while the dominant force can be assumed to be gravity alone, acting in a downwards direction. Setting t for time, we have:

$$m\frac{d^2x}{dt^2} = 0,$$

$$m\frac{d^2y}{dt^2} = -mg,$$

where m is the mass of the shot (assumed constant) and g stands for the acceleration due to gravity (on earth, about 9.8 metres per second per second). Air resistance is ignored at this stage, and the force due to gravity is an approximation valid near to the earth's surface; a more accurate treatment is discussed in Section 11.4.

Since y is measured positive upward while gravity acts downward, its contribution comes with a negative sign in the second equation.

At this level of approximation, the mass of the shot is actually irrelevant: it has no effect in the first equation and cancels out in the second equation. So,

$$\frac{d^2x}{dt^2} = 0, \qquad (11.1)$$

$$\frac{d^2y}{dt^2} = -g. \qquad (11.2)$$

Assume now that at $t = 0$, both x and y are zero, meaning that the cannon is on the ground and that horizontal distances are measured from the cannon position. Assume also that the cannon is pointing upward at an angle θ to the horizontal, and muzzle speed is v. Then, using the trigonometric equations from Chapter 3, the initial horizontal velocity component is $v \cos \theta$ and the initial vertical component is $v \sin \theta$.

Integrate equations (11.1) and (11.2) once with respect to t, keeping g constant, to get the speed components:

$$\frac{dx}{dt} = a,$$

$$\frac{dy}{dt} = -gt + b.$$

The inclusion of the constants a and b allow for what might well be present but which disappear during the differentiation.

Applying the initial speeds at $t = 0$ gives $a = v \cos \theta$ and $b = v \sin \theta$, so that:

$$\frac{dx}{dt} = v \cos \theta, \qquad (11.3)$$

$$\frac{dy}{dt} = -gt + v \sin \theta. \qquad (11.4)$$

11.1. FLIGHT OF A CANNONBALL

Integrate again with respect to t, keeping v and θ constant:

$$x = vt\cos\theta + c,$$
$$y = -\frac{1}{2}gt^2 + vt\sin\theta + d,$$

introducing two new constants, c and d.

If, as assumed, $x = y = 0$ at $t = 0$, then $c = d = 0$ and the final trajectory is given by:

$$x = vt\cos\theta, \tag{11.5}$$
$$y = -\frac{1}{2}gt^2 + vt\sin\theta. \tag{11.6}$$

A look at equation (11.6) shows that y is zero both at $t = 0$ and when $-gt/2 + v\sin\theta = 0$, and this latter equation gives the flight time t_f of the shot:

$$t_f = \frac{2v\sin\theta}{g}.$$

Substituting this into equation (11.5) enables determination of the impact distance relative to the cannon — the ground range x_f:

$$x_f = \frac{v^2\sin 2\theta}{g}, \tag{11.7}$$

using the relation $2\sin\theta\cos\theta = \sin 2\theta$ (equation (3.7)). From this, if x_f is measured and θ was recorded at the outset, the muzzle speed is obtainable:

$$v = \sqrt{\frac{g\,x_f}{\sin 2\theta}}.$$

Add some figures to this. For the small brass cannon, assume a ground range of about 4 metres and $\theta = 30°$; then v works out at about 6.7 metres per second — quite a respectable speed for a tiddly cannon.

Equation (11.7) will also enable the prediction of the *maximum possible* ground range for the same v, by varying θ. Since $\sin 2\theta$ has its maximum value of unity at $2\theta = 90°$, pointing the cannon up at $45°$ will project the shot the furthest. This can also be seen from Figure 11.1, in which the trajectories for various value of θ are plotted.

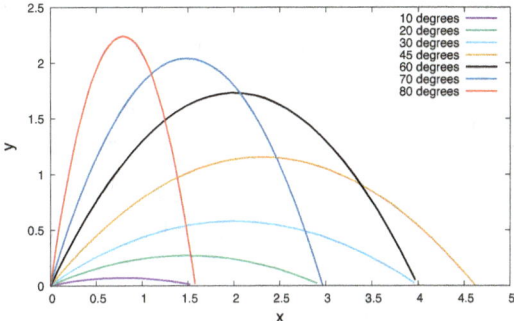

Figure 11.1: Example cannon shot trajectories

It may also be observed that ranges less than the maximum can be reached using two different elevation angles. The higher-altitude lofted trajectories are used by military mortars.

Thus far, air resistance has been neglected. From aerodynamic considerations, the drag force is proportional to the speed squared, to the air density ρ, to the area A of the projectile and inversely to the mass, so the kinematic equations of motion are:

$$\frac{d^2x}{dt^2} = -\frac{C_D A \rho}{2m}\frac{dx}{dt}\sqrt{\left(\frac{dx}{dt}\right)^2 + \left(\frac{dy}{dt}\right)^2},$$

$$\frac{d^2y}{dt^2} = -g - \frac{C_D A \rho}{2m}\frac{dy}{dt}\sqrt{\left(\frac{dx}{dt}\right)^2 + \left(\frac{dy}{dt}\right)^2}.$$

The square root quantity arises as a consequence of decomposing the net drag force, which is proportional to $(dx/dt)^2 + (dy/dt)^2$, into components in the x and y directions. C_D is the *drag coefficient*, which is a function of shape (pointy projectiles have smaller C_D and so less drag). With air resistance included, the mass of the projectile now becomes relevant, and the heavier the shot, the less the resistance.

These equations need to be solved numerically.

The two equations can also be used to estimate the total force on a fence panel (for instance), when under wind pressure; wind moving against a static panel is dynamically the same as the panel moving and the air stationary. So assume motion in the x-axis only and form the force F as mass × acceleration:

$$F = \frac{1}{2}C_D A \rho v^2,$$

since the mass cancels out and where $v^2 = (dx/dt)^2$ is the air speed squared. For a flat square sheet, C_D is approximately unity [37] (assuming no gaps or holes) and the air density at the earth's surface is about 1.22 kilograms per cubic metre, so we get the following approximation

$$F \approx 0.61 A v^2.$$

The force is thus proportional to the area and to the *square* of the wind speed, which helps explain why tornados and hurricanes are so destructive.

11.2 Pendulum Motion

Old-fashioned pendulum clocks are well-known for keeping good time (except at sea or on other unstable platforms), but what is it about this particular control mechanism that makes it reliable? Newton's law of motion again helps in understanding.

Suppose that the pendulum consists of a heavy, compact mass m suspended from a pivot using a thin stiff rod of fixed length l, the aim being to minimise air

11.2. PENDULUM MOTION

resistance during the motion. In Figure 11.2, x is the horizontal displacement of the pendulum bob, while y is its vertical distance below the pivot line.

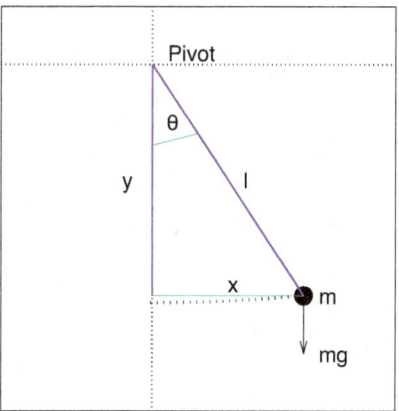

Figure 11.2: Pendulum coordinates

Applying Newton's law in the two individual coordinate directions:

$$m\frac{d^2x}{dt^2} = 0, \tag{11.8}$$

$$m\frac{d^2y}{dt^2} = mg, \tag{11.9}$$

where t stands for time. Since y is measured *downwards* from the pivot, the gravitational force acts in the same direction.

With reference to Chapter 3), $x = l\sin\theta$ and $y = l\cos\theta$, but l is treated as a constant and only θ varies. However, θ itself is a function of time, so we need the equation:

$$\frac{dx}{dt} = \frac{dx}{d\theta}\frac{d\theta}{dt}.$$

Regarding the $d\theta$, etc., as small quantities, one can mentally 'cancel' the $d\theta$ quantity and see that this equation does make sense.

Therefore,

$$\frac{dx}{dt} = l\cos\theta\frac{d\theta}{dt}, \tag{11.10}$$

$$\frac{dy}{dt} = -l\sin\theta\frac{d\theta}{dt}, \tag{11.11}$$

To differentiate again, a further rule is needed for the product of two functions. Suppose $z = u(t)v(t)$, where both u and v are functions of time. Then,

$$\frac{dz}{dt} = u(t)\frac{dv}{dt} + \frac{du}{dt}v(t).$$

With this in mind, differentiating equations (11.10) and (11.11) once more with respect to time and incorporating the results into equations (11.8) and (11.9), gives the following kinematic equations:

$$ml\left[\cos\theta\frac{d^2\theta}{dt^2} - \sin\theta\left(\frac{d\theta}{dt}\right)^2\right] = 0,$$

$$ml\left[-\sin\theta\frac{d^2\theta}{dt^2} - \cos\theta\left(\frac{d\theta}{dt}\right)^2\right] = mg.$$

In both cases, the mass m cancels out, and l can be dropped in the first equation, so

$$\cos\theta\frac{d^2\theta}{dt^2} - \sin\theta\left(\frac{d\theta}{dt}\right)^2 = 0,$$

$$-\sin\theta\frac{d^2\theta}{dt^2} - \cos\theta\left(\frac{d\theta}{dt}\right)^2 = \frac{g}{l}.$$

If the first of these is rearranged to get

$$\left(\frac{d\theta}{dt}\right)^2 = \frac{\cos\theta}{\sin\theta}\frac{d^2\theta}{dt^2},$$

this can be substituted into the second, resulting in:

$$-\sin\theta\frac{d^2\theta}{dt^2} - \cos\theta\frac{\cos\theta}{\sin\theta}\frac{d^2\theta}{dt^2} = \frac{g}{l}.$$

Multiplying both sides by $-\sin\theta$ and recalling that $\cos^2\theta + \sin^2\theta = 1$ (from Chapter 3), we end up with:

$$\frac{d^2\theta}{dt^2} = -\frac{g}{l}\sin\theta. \tag{11.12}$$

Suppose now that the pendulum only swings over a small arc, so θ remains small. Then, from the power series expansions in Section 7.3, the kinematic equation in θ is approximated by:

$$\frac{d^2\theta}{dt^2} \approx -\frac{g}{l}\theta. \tag{11.13}$$

This is satisfied either by $\sin\left(t\sqrt{g/l}\right)$ or $\cos\left(t\sqrt{g/l}\right)$, implying that the pendulum period T will be:

$$T = 2\pi\sqrt{\frac{l}{g}} \quad \text{(obtainable from } T\sqrt{g/l} = 2\pi\text{).}$$

This is the time taken for one complete forward and back pendulum swing. It depends only on the local acceleration due to gravity and on the length of the pendulum strut — the mass is irrelevant. The longer the pendulum strut, the slower the clock tick.

With our old long-case clock, the period is two seconds (so far as can be determined using the stopwatch on a mobile phone). This is the forward and back period, whereas the actual audible 'tick' occurs at each end of the pendulum swing, thus giving the one-second quantisation. To obtain $T = 2$ seconds, the length l needs to be one metre long, almost exactly[3].

The exact equation (11.12) can be solved in terms of elliptic functions, which are a sort of generalised sines and cosines, and also periodic in nature. The equation can also be solved numerically and compared to the simpler solutions for equation (11.13). It turns out that the exact solution has a slightly longer period than does the sinusoidal approximation (the analysis is fun to do but a little more complex and, somewhat reluctantly, has been hived off to Appendix C).

11.3 Rocket Thrust

Applying Newton's equation, force = mass × acceleration, requires some description of the force, which is straightforward enough when working with gravity or a human-applied pressure, for instance. But what about rockets? These gain thrust by ejecting high-speed gas in the *opposite* direction to rocket travel, so how does this work? It is not a matter of the gas 'pushing against' the atmosphere, since rockets work perfectly well in space with no atmosphere at all.

The underlying principle is that of conservation of momentum, momentum being defined as mass × velocity [9]. Suppose at some time t the mass of the rocket is m and its velocity is v, heading to the right. A short time dt later, a small amount dm of the rocket's fuel has been burned and was ejected at a speed c to the left. So the momentum at time t is given by mv, while at time $t + dt$ the momentum of the complete system (rocket plus burned fuel) is:

$$(m - dm)(v + dv) - (c - v)\,dm.$$

This needs a bit of explanation. The first term here, $(m - dm)(v + dv)$, is the new momentum of the rocket, which has lost some mass but gained in speed. The second term, $-(c - v)\,dm$, is the momentum of the burned rocket fuel, the mass being dm and the speed being $c - v$. The negative sign occurs because the gas was ejected to the left, while the rocket moves to the right, and the quantity $c - v$ appears rather than just c because the rocket was moving at the same time as burning the fuel, and in the opposite direction[4].

---

[3]The acceleration due to gravity, g, varies slightly from place to place. Pendulum clocks generally have fine adjustment of l, to allow for this as well as for temperature variations and the other approximations.

[4]To get a sense of this, consider an aircraft firing a missile forward. The missile will then have its own speed relative to the ground augmented by that of the launching aircraft. Firing backward would lose missile speed and could even cancel it out (relative to the ground) if the aircraft was fast enough. This simple logic needs amendment in relativity, though; see Chapter 28.

Now equate the two momenta at time t and $t + dt$:

$$mv = (m - dm)(v + dv) - (c - v)\,dm,$$
$$= mv + m\,dv - v\,dm - dm\,dv - c\,dm + v\,dm, \quad \text{on expanding out.}$$

After cancelling terms, this becomes:

$$m\,dv - dm\,dv - c\,dm = 0.$$

Divide this by dt and let dt, dv and dm all go to to zero. The result is, simply,

$$m\frac{dv}{dt} = c\frac{dm}{dt}.$$

The left-hand side is mass times acceleration, while the right-hand-side defines the effective force pushing the rocket on its way. The exhaust gas velocity c is typically quite large, so the thrust can be large also.

11.4 Planetary Orbits

Newton provided an equation for the gravitational force between two objects:

$$F = \frac{GMm}{r^2}, \tag{11.14}$$

where G is the universal gravitational constant, M and m are the masses of the objects and r is the distance between their centres. This attractive force acts along the line joining the two objects. If M is the earth's mass and the dynamics of interest takes place near to the earth surface, then the quantity $GM/r^2 \approx g$, thus giving $F = mg$ as in section 11.1. This approximation works because the equatorial radius r of the earth, at 6378137 metres, is very much larger than small altitudes relative to the earth's surface, which can be ignored in comparison.

As an aside, for a spherical mass of uniform constitution, the combined gravitational effect of all the individual mass points in the earth is as if the entire mass was concentrated at the centre; this is why we are dealing with a single value of r rather than an integral over a spherical volume. For more accurate work, the oblateness of the earth needs to be taken into account (it is squished in a bit at the poles, besides being lumpy), and this asphericity can have profound long-term effects on earth-orbiting satellites.

But enough of the diversion — how can we work out from equation (11.14) how planets move around the sun, or satellites move round the earth? Newton managed it more than 300 years ago.

To simplify, we assume that the orbit takes place in a plane, so only two spatial coordinates are needed. The presence of other objects (planets, *etc*) complicate the shapes of real orbits, but we concentrate here on only two masses. So in the spatial coordinates x and y of interest, centred on the much larger

11.4. PLANETARY ORBITS

mass $M \gg m$,

$$m\frac{d^2x}{dt^2} = -\frac{GMm}{r^2}n_x,$$

$$m\frac{d^2y}{dt^2} = -\frac{GMm}{r^2}n_y,$$

the left-hand-sides being mass × acceleration. The quantities n_x and n_y stand for components of direction, such that

$$n_x = \frac{x}{r} \text{ and } n_y = \frac{y}{r},$$

with $n_x^2 + n_y^2 = 1$. This use of vector components n_x, n_y anticipates the discussion in Chapter 14.

It can be seen in the above equations that m occurs on both sides of both equations and so cancels out. Therefore,

$$\frac{d^2x}{dt^2} = -\frac{GM}{r^3}x, \tag{11.15}$$

$$\frac{d^2y}{dt^2} = -\frac{GM}{r^3}y, \tag{11.16}$$

with $r = \sqrt{x^2 + y^2}$ (Pythagoras again).

Instead of working in the Cartesian[5] coordinates x and y, it is more convenient to adopt polar coordinates r and θ, r being as above and θ measuring an angle in the orbit plane. So set

$$x = r\cos\theta, \quad y = r\sin\theta.$$

Differentiate these with respect to time, bearing in mind the rules for differentiating products of variables (see Section 7.10):

$$\frac{dx}{dt} = \frac{dr}{dt}\cos\theta - r\frac{d\theta}{dt}\sin\theta,$$

$$\frac{dy}{dt} = \frac{dr}{dt}\sin\theta + r\frac{d\theta}{dt}\cos\theta.$$

And differentiate once more:

$$\frac{d^2x}{dt^2} = \frac{d^2r}{dt^2}\cos\theta - 2\frac{dr}{dt}\frac{d\theta}{dt}\sin\theta - r\frac{d^2\theta}{dt^2}\sin\theta - r\left(\frac{d\theta}{dt}\right)^2\cos\theta,$$

$$= \cos\theta\left[\frac{d^2r}{dt^2} - r\left(\frac{d\theta}{dt}\right)^2\right] - \sin\theta\left[2\frac{dr}{dt}\frac{d\theta}{dt} + r\frac{d^2\theta}{dt^2}\right],$$

(collecting sine and cosine terms),

$$\frac{d^2y}{dt^2} = \frac{d^2r}{dt^2}\sin\theta + 2\frac{dr}{dt}\frac{d\theta}{dt}\cos\theta + r\frac{d^2\theta}{dt^2}\cos\theta - r\left(\frac{d\theta}{dt}\right)^2\sin\theta,$$

$$= \cos\theta\left[2\frac{dr}{dt}\frac{d\theta}{dt} + r\frac{d^2\theta}{dt^2}\right] + \sin\theta\left[\frac{d^2r}{dt^2} - r\left(\frac{d\theta}{dt}\right)^2\right].$$

---

[5]Named after René Descartes, 1596 to 1650.

Now plonk these into the left-hand-sides of equations (11.15) and (11.16):

$$\cos\theta\left[\frac{d^2r}{dt^2} - r\left(\frac{d\theta}{dt}\right)^2\right] - \sin\theta\left[2\frac{dr}{dt}\frac{d\theta}{dt} + r\frac{d^2\theta}{dt^2}\right] = -\frac{GM}{r^2}\cos\theta, \quad (11.17)$$

$$\cos\theta\left[2\frac{dr}{dt}\frac{d\theta}{dt} + r\frac{d^2\theta}{dt^2}\right] + \sin\theta\left[\frac{d^2r}{dt^2} - r\left(\frac{d\theta}{dt}\right)^2\right] = -\frac{GM}{r^2}\sin\theta, \quad (11.18)$$

here using the above definitions of x and y in terms of r and θ.

Multiply the first of these throughout by $\cos\theta$ and the second throughout by $\sin\theta$ and add the two together. Using the known relation $\cos^2\theta + \sin^2\theta = 1$, we find that:

$$\frac{d^2r}{dt^2} - r\left(\frac{d\theta}{dt}\right)^2 = -\frac{GM}{r^2}. \quad (11.19)$$

Substituting this back into either of equations (11.17) or (11.18) results in:

$$2\frac{dr}{dt}\frac{d\theta}{dt} + r\frac{d^2\theta}{dt^2} = 0. \quad (11.20)$$

Now, if $r \neq 0$ (which hopefully won't occur for a practical orbit), equation (11.20) can be multiplied throughout by r, in which case

$$\frac{d}{dt}\left[r^2\frac{d\theta}{dt}\right] = 0, \text{ which implies that } r^2\frac{d\theta}{dt} = h, \quad (11.21)$$

where h is a constant.

Using this to plug $d\theta/dt$ back into equation (11.19) then results in a single equation in r:

$$\frac{d^2r}{dt^2} - \frac{h^2}{r^3} = -\frac{GM}{r^2}. \quad (11.22)$$

An exact solution of this can be found if, instead of working with dr/dt, we look at $dr/d\theta$ instead. So,

$$\frac{dr}{dt} = \frac{dr}{d\theta}\frac{d\theta}{dt} = \frac{h}{r^2}\frac{dr}{d\theta}, \text{ from equation (11.21)}.$$

Differentiate this with respect to t, again changing from t to θ:

$$\frac{d^2r}{dt^2} = \frac{d}{dt}\left(\frac{dr}{dt}\right) = \frac{d}{d\theta}\left(\frac{dr}{dt}\right)\frac{d\theta}{dt},$$

$$= \frac{d}{d\theta}\left(\frac{h}{r^2}\frac{dr}{d\theta}\right)\frac{d\theta}{dt} = \frac{h}{r^2}\frac{d}{d\theta}\left(\frac{h}{r^2}\frac{dr}{d\theta}\right),$$

$$= \frac{h^2}{r^2}\frac{d}{d\theta}\left(\frac{1}{r^2}\frac{dr}{d\theta}\right), \text{ since } h \text{ is a constant}.$$

11.4. PLANETARY ORBITS

Therefore, using this in equation (11.22), we get:

$$\frac{d}{d\theta}\left(\frac{1}{r^2}\frac{dr}{d\theta}\right) - \frac{1}{r} = -\frac{GM}{h^2},$$

after dividing throughout by h^2 and cancelling the common r^2 term. It might appear from this that we are as far away as ever from a compact solution, but the first bracketed term can also be written in the form:

$$\frac{d}{d\theta}\left[-\frac{d}{d\theta}\left(\frac{1}{r}\right)\right] - \frac{1}{r} = -\frac{GM}{h^2}.$$

So if we define $u = 1/r$, we get

$$\frac{d^2 u}{d\theta^2} + u = \frac{GM}{h^2},$$

and u must then have sinusoidal solutions. That is,

$$u = A\cos\theta + B\sin\theta + \frac{GM}{h^2},$$

where A and B are constants. Reverting to r and writing $A\cos\theta + B\sin\theta = C\cos(\theta - \theta_0)$ then results in:

$$r = \frac{1}{C\cos(\theta - \theta_0) + GM/h^2},$$
$$= \frac{h^2/GM}{1 + C'\cos(\theta - \theta_0)}, \text{ where } C' = \frac{Ch^2}{GM}.$$

The quantity C' is called the *eccentricity* of the curve, and its value characterises a family of curve types, of which the ellipse is one example. For instance, if $C' = 0$, we end up with a circle (a special case of an ellipse), with $r = $ constant. The easiest way of seeing the types of curve produced is to generate a graph with a range of C' values, as in Figure 11.3.

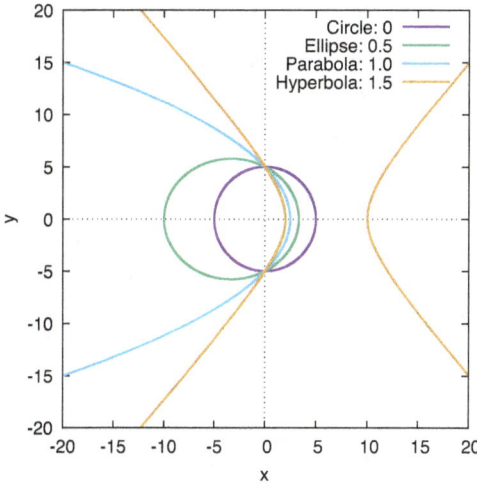

Figure 11.3: Conic section curves

Here, the quantity h^2/GM is set quite arbitrarily to a value of 5, merely in order to generate the picture. With $C' = 0$, a circle is produced (as expected), while $C' = 0.5$ gives an ellipse, $C' = 1$ a parabola and $C' = 1.5$ a hyperbola (which has two branches, as shown)[6]. All of these are called 'conic section' curves, since they can be obtained by cutting a solid cone at various angles.

To provide a bit of historical background, Kepler[7] deduced that the planetary orbits must be ellipses on the basis of the meticulous astronomical observations carried out by Brahe[8]. But it was Newton who was able to show, using calculus, that an inverse-square gravitational law must result in conic section curves.

---

[6] More generally, $|C'| < 1$ produces an ellipse, $|C'| = 1$ a parabola and $|C'| > 1$ a hyperbola.
[7] Johannes Kepler, 1571 to 1630.
[8] Tycho Brahe, 1546 to 1601.

Chapter 12

MOMENTS AND BALANCE

SUPPOSE we have a book located near to the edge of the table. Your cat[1] then pushes it progressively toward the edge until it over-hangs; at what point will it finally topple off? Intuitively (and from experience), the book will need to be more than half over the edge before it falls (at least if the book is of a uniform construction).

A useful rule-of thumb is to see where the book's centre of mass is in relation to the table edge: if over the edge, the book will fall, but otherwise will stay where it is[2]. The centre of mass is a very useful concept, but in the present context the idea of a *moment* takes us further. The word 'moment' here has nothing to do with time increments, but instead describes a turning force — a torque.

The idea of a torque is most easily explained in terms of two weights on a beam, as in Figure 12.1.

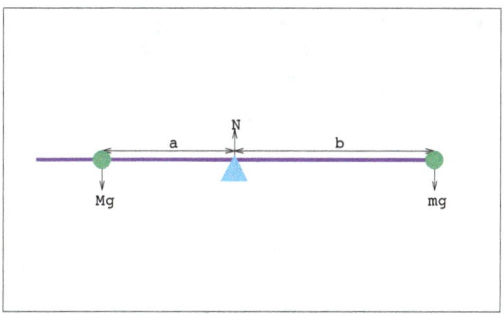

Figure 12.1: Balance example

---

[1] A small furry animal with whiskers.
[2] At least until the cat gives it another nudge.

CHAPTER 12. MOMENTS AND BALANCE

The beam itself is here assumed to be light, or at least much lighter than the two masses M and m that are illustrated, and also extremely stiff so that it does not bend. These masses are of different magnitude and at different distances from the pivot point, so the beam as a whole is balanced and remains horizontal. The quantity g stands for the local acceleration due to gravity and is treated as constant (see Section 11.1) above).

The quantity N pointing upwards from the pivot point stands for the *reaction force*. From Newton's law of action and reaction, the pivot has to be pressing upwards to counteract the two weights pushing downwards, since the whole structure is static. The concept of reaction will be needed when we return to the over-balancing book situation and in bending beam engineering problems.

In Figure 12.1, the left-hand mass contributes a downward force of Mg and is situated a distance a from the pivot point. Its contribution to the torque will then be Mga, acting in an anti-clockwise manner. That is, the torque is the force multiplied by the perpendicular distance to the pivot or axis of rotation[3].

In contrast, the right hand mass provides a torque of mgb, acting clockwise about the same pivot. Since the beam (and superimposed weights) are in equilibrium and the beam remains horizontal, the two torques must cancel each other out, so that

$$Mga = mgb, \text{ which simplifies to } Ma = mb,$$

since g occurs on both sides of the equation. This equation is the basis of the *steelyard*, in which a quantity to be weighed is balanced by one or more known masses that can be shifted left or right along the bar until the latter is horizontal.

An obvious corollary is that if $a = b$, then the two masses must be the same if the beam is to remain horizontal.

Now return to the book on the edge of the table, as illustrated in Figure 12.2. Here, the table is coloured purple and the book (with an appropriately whimsical title) is in green.

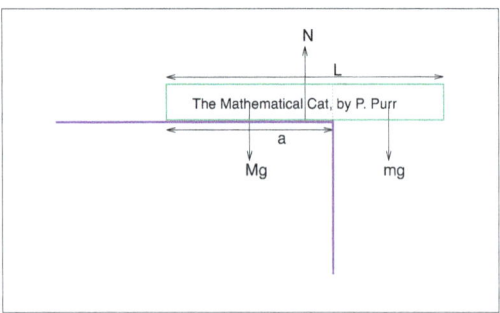

Figure 12.2: Book balancing example

The book is of length L and positioned such that most of its bulk is still fully

---

[3]More generally, the torque vector T is given by $\mathbf{T} = \mathbf{r} \wedge \mathbf{N}$, where the $\wedge$ symbol stands for the vector product and $\mathbf{r}$ is the vector from $\mathbf{N}$ to the pivot [6].

on the table top, with its left-hand end at a distance a from the table edge. The amount of book over-hang is, therefore, $L - a$.

The dominant mass of the book, amounting to M for the on-table part, has a centre of mass half-way along a, so its torque will be $Mga/2$, acting in an anti-clockwise direction around the table edge (the pivot point). Similarly, the book over-hanging section has a smaller mass m and its torque will be $mg(L-a)/2$ and acting in a clockwise direction around the edge.

The final contribution to the set of torques is from the reaction force N, acting through the book's centre of mass which will be at a distance $a - L/2$ from the edge. This torque will also act in a clockwise direction. So, since the book is currently stable, the set of moments must balance, giving the equation:

$$\frac{Mga}{2} = N\left(a - \frac{L}{2}\right) + \frac{mg(L-a)}{2}. \tag{12.1}$$

We also know that the reaction force N must also balance the total mass of the book, giving $N = (M+m)g$ (or else the book would fall through the table). So the equation actually involves only masses and lengths, since g cancels out.

Now imagine progressively pushing the book to the right, to a point where N points upward directly above the table edge. At this point, N ceases to contribute any torque and drops out of equation (12.1), giving

$$Ma = m(L-a),$$

after cancelling the common factor $g/2$. The condition for the book falling off the table then comes down to whether the clockwise turning force from mg is larger than the anti-clockwise Mg torque, in accord with intuition (and experiment).

12.1 Ladder Against a Wall

Most of us who have needed to go up a ladder for one purpose or other are likely to have had some concerns over safety. And most of us, whether confident with ladders or not, will have appreciated (instinctively or otherwise) the vital importance of frictional forces in this context. So it is worth examining this situation from a mathematical point of view.

This is a useful example of a problem where, in the general case, a complete solution cannot be obtained, requiring the invocation of additional physics-based criteria to break the deadlock. The manipulation of the various inequalities involved is also instructive, since not much progress can be made without their inclusion.

Refer to Figure 12.3, which shows a schematic of a ladder resting against a wall and with its feet on the ground.

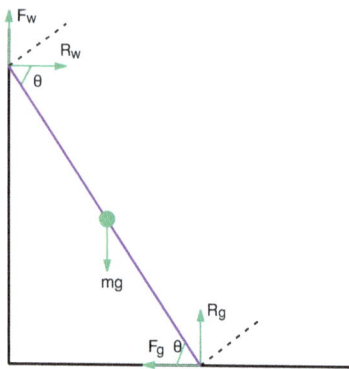

Figure 12.3: Forces holding a ladder against a wall

For simplicity, this is the ladder without anyone attempting to go up it, which is sufficient to bring in the key concepts. The quantity mg pointing downwards represents the weight of the ladder, for mathematical purposes concentrated at its centre. The R_g and R_w quantities illustrated are *reaction* forces; from Newton's law of action and reaction, these are a consequence of the ground holding up the ladder and the wall preventing its top end from falling through.

The other two quantities, F_g and F_w, are frictional forces. These are needed to prevent slippage: just imagine what would happen if the wall on the left was made of glass, and the ground was a sheet of ice — experience says that in no time at all, the ladder would be lying flat. Continuing with this 'thought experiment', it is evident that such a ladder collapse would involve a rotation as well as a general shift to the right. So the frictional forces combined with the reaction forces must be exerting a moment, or torque, on the ladder, a concept that was introduced in the previous section.

But before looking a bit more closely at the mathematical treatment of friction, let us first see what can be gained from resolving the net forces on the ladder in the horizontal x and vertical y directions:

In the x direction: $R_w - F_g = 0$, (12.2)

In the y direction: $R_g + F_w - mg = 0$. (12.3)

The only quantity that is known here is mg, so there are four unknowns and only two equations, meaning that the various unknown forces cannot be determined uniquely.

Add in the torque component to see if this helps, and use the ladder centre of mass as the pivot point. Using the ladder angle θ to resolve the forces in directions perpendicular to the ladder length[4], we get:

In the clockwise direction: $R_w \sin\theta + F_w \cos\theta + F_g \sin\theta$,

In the anti-clockwise direction: $R_g \cos\theta$.

---

[4]The dashed lines in Figure 12.3 are normal to the ladder and are intended to aid the mental vector resolving process.

12.1. LADDER AGAINST A WALL

Since the ladder is not rotating, these two equations can be equated. Then, dividing throughout by $\cos\theta$ and rearranging ends up with:

$$(R_w + F_g)\tan\theta = (R_g - F_w). \tag{12.4}$$

There are now three equations but still four unknowns, so the solution remains indeterminate.

Now have a look at the friction side of things. From intuition — and experiments — it is usual to assume that a frictional force is proportional to the related reaction force [38],[39], so that $F_w \propto R_w$ and $F_g \propto R_g$. This concords with experience: imagine a largish stone resting on a flat (but not smooth) surface. Attempting to push the stone along the surface will take a lot more effort for a heavy stone than for a lighter one.

In fact, if the pushing process is made more experimentally rigorous by steadily increasing the applied force until slippage occurs, while recording this force at each time step, we would end up with something akin to Figure 12.4.

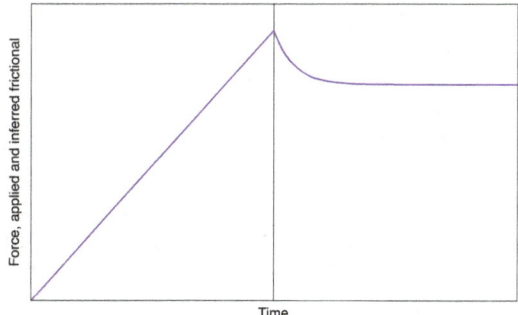

Figure 12.4: Applied force versus time

This behaviour reflects experience: starting from scratch and pushing progressively harder initially yields no results, until the stone abruptly gives way and then slides with less effort required than to get it moving. Figure 12.4 does now provide some insight into how to model friction in mathematical terms. Unfortunately, unequivocal measurement of the frictional force only occurs once the stone starts to slip and thereafter, since only then does the applied force equal the friction force. So it is possible to measure F_c — the peak, or critical, value — and relate this to the normal reaction force R, so that

$$F_c = \mu_c R, \tag{12.5}$$

where μ_c is a derived proportionality constant. Reference [40] provides values for various material combinations.

The kinetic friction régime to the right of the critical point in Figure 12.4 requires different considerations and is not further examined here.

Prior to the critical time, all one can really say is that the friction force must be less than F_c, implying that a more general relation must be of the form

$$F \leq \mu_c R. \tag{12.6}$$

This just states that, for stability, the applied load cannot be permitted to be large enough to overcome F_c.

A frequently-applied simplification that can be made at this point is to assume a smooth wall, so that $F_w = 0$. Therefore, equations (12.2), (12.3) and (12.4) become:

$$R_w = F_g,$$
$$R_g = mg,$$
$$(R_w + F_g)\tan\theta = R_g.$$

Substituting the first into the last, we get:

$$2F_g \tan\theta = R_g, \text{ which implies that } F_g = \frac{R_g}{2\tan\theta}.$$

Using $F_g \leq \mu_g R_g$ from the above frictional inequalities, then for the ladder not to slip[5]:

$$\tan\theta \geq \frac{1}{2\mu_g}, \tag{12.7}$$

which is a familiar stability inequality for a ladder leaning against a smooth wall. Intuitively, putting the foot of the ladder closer to the wall should result in a more stable situation, since then its weight will then pass more directly downward to the ground.

An alternative scenario is where it is assumed that *both* surfaces slip simultaneously, giving $F_w = \mu_w R_w$ and $F_w = \mu_w R_w$. If these prescriptions are taken literally, then equations (12.2), (12.3) and (12.4) provide a set of three equations in only two unknown quantities (namely R_g and R_w). Contradictions then arise — *except* where θ is constrained to be the following specific function of the two friction coefficients:

$$\tan\theta = \frac{1 - \mu_w \mu_g}{2\mu_g}.$$

Thus, if θ happens to be set to this particular value, then all three equations are consistent and a unique solution is possible. In the more general case, with a user-chosen choice of θ, one may conclude that both top and bottom of the ladder do not slip at the same time.

The case where $F_w \neq 0$ is more complicated. One can envisage a situation where slippage is allowed at the wall, but sufficient friction still pertains on the ground to prevent the ladder from moving[6]. This is the example examined in [39],[41], whereby the friction at the wall is insufficient to prevent ladder collapse if the ground is smooth or nearly so. So for a rough ground and a smoother wall, it can be assumed that $F_w = \mu_w R_w$ — a near-slip constraint[7].

---

[5] In rearranging this inequality, it is assumed that $\tan\theta$ is positive.

[6] Personally, I'd be a bit worried that the top of the ladder may then slip sideways on the wall ...

[7] With a person ascending the ladder, additional physics-based considerations can be imported to aid the analysis. When a person climbs up, the ladder itself is expected to flex under the additional load, whereupon the top of the ladder slips a bit downwards before managing to grip again.

12.2. BEAM EQUATIONS

From equations (12.2) and (12.4), plus using the equality for F_w, we get:

$$2F_g \tan\theta = R_g - \mu_w R_w,$$
$$= R_g - \mu_w F_g.$$

Rearrange this to get:

$$F_g = \frac{R_g}{2\tan\theta + \mu_w}.$$

Apply the above inequality on F_g, resulting in

$$\frac{R_g}{2\tan\theta + \mu_w} \leq \mu_g R_g.$$

This can then be rearranged to give an inequality for $\tan\theta$:

$$\tan\theta \geq \frac{1}{2\mu_g}\left(1 - \mu_w \mu_g\right).$$

If $\mu_w = 0$, this reduces to equation (12.7).

12.2 Beam Equations

We now turn to situations a bit closer to engineering practice, in which a beam is under one or more loads. A simple situation is illustrated in Figure 12.5.

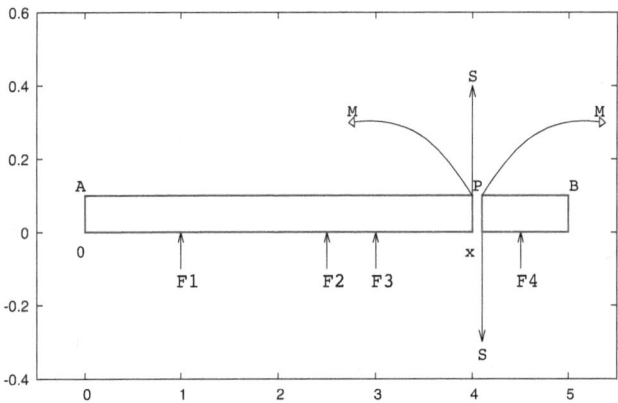

Figure 12.5: Discrete loads on a beam

For the time being, ignore the gap in the diagram and assume that the beam AB is in equilibrium under several upward forces, here labelled F_1 to F_4. Since the beam isn't going anywhere, the applied forces must cancel out, so that

$$\sum_{i=1}^{4} F_i = 0.$$

It can be appreciated that at least some of the forces must be pointing downwards for this equation to make physical sense, but this realisation can be accommodated simply by allowing negative signs where appropriate.

In a similar manner, taking moments of the forces about A results in:

$$\sum_{i=1}^{4} x_i F_i = 0,$$

measuring distances x_i from A.

And if the beam happens to be pinned into the wall at A, as with a cantilever, there will be an additional moment preventing the left-hand end from turning under the other loads.

Return now to Figure 12.5 and choose some arbitrary point P at a distance x from A. The section AP is in equilibrium because the segment PB is exerting a *shearing force* S on AP at P. Similarly, the section PB is in equilibrium because AP exerts a shearing force S is the opposite direction. These shearing forces are a consequence of parts of the beam holding other parts up, the whole assembly being static. By convention, S is assigned a positive value if it acts upwards on the left of a cross-section at P.

In a similar manner, PB exerts a couple on AP known as the *bending moment* M at P, and PA exerts an equal and opposite couple on PB. Also by convention, M is taken as positive if it acts anti-clockwise on the left of a cross-section at P.

Resolving the forces for the AP segment of the beam, we must have, for equilibrium:

$$S = -\sum_{i=1}^{3} F_i;$$

excluding F_4 since this is to the right of P.

Similarly, taking moments about P,

$$M = \sum_{i=1}^{3} (x - x_i) F_i.$$

This can be understood if we regard P as a pivot point; each of the F_1, F_2, F_3 then act to turn the section AP clockwise about P, with each at a different distance from P. The moment M then needs to act in an anti-clockwise direction to balance things up, as shown in Figure 12.5 and with a positive sign.

Looking back at Figure 12.5, commencing x at zero and then moving it progressively to the right, it can be appreciated that S will undergo abrupt changes as each of the F_i is passed. So the shear S may be *discontinuous* in x, depending on the nature of the applied load or loads. On the other hand, M will be continuous but potentially will have discontinuous derivatives.

Consider now situations in which the beam (or the book, for that matter) starts to bend under a continuous rather than discrete load, and refer to Figure 12.6, a variation on Figure 12.5.

12.2. BEAM EQUATIONS

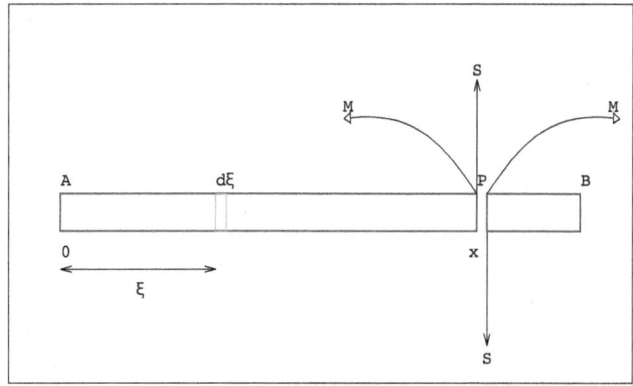

Figure 12.6: Continuous forces on a beam

The beam AB is now held horizontally in equilibrium under the action of some external *downward* force per unit length $f(\xi)$ (such as gravity), where ξ is measured from A. Choose some arbitrary point P along the beam, at distance x from A. As before, the section AP is in equilibrium due to segment PB exerting a shearing force S on AP at P. Similarly, the section PB is in equilibrium due to AP exerting S in the opposite direction.

In the same manner, the right-hand section PB exerts a torque M on the left-hand section AP, with the turning point at A. This is balanced by the same torque M acting on section PB.

Consider now some element of length $d\xi$ located at distance ξ from A. The downward force on the element $d\xi$ will then be $f(\xi)d\xi$, so to achieve equilibrium at P the cumulative downward force at P must be balanced by the upward shear force, giving [33]

$$S = \int_0^x f(\xi)d\xi. \tag{12.8}$$

Take moments about P, so that

$$M = -\int_0^x (x-\xi)f(\xi)d\xi,$$
$$= \int_0^x \xi f(\xi)d\xi - xS, \text{ after some simplification.} \tag{12.9}$$

Differentiate equation (12.8) with respect to x, so that

$$\frac{dS}{dx} = f(x), \tag{12.10}$$

thus relating the rate of change of shear force with x to the external force per unit length.

Similarly, differentiate equation (12.9) with respect to x:

$$\frac{dM}{dx} = xf(x) - S - x\frac{dS}{dx},$$
$$= -S, \text{ from equation (12.10).}$$

Or, differentiating again,

$$\frac{d^2M}{dx^2} = -f(x). \tag{12.11}$$

Equations (12.10) and (12.11) relate the shear force and the torque to the external force, but a bit more analysis is needed to bring in the actual beam curvature under the influence of f (which could be gravity or some other load). As before, x is measured horizontally from A, while y is measured positive downwards on the assumption that the beam is bending in a downwards direction. The effects of the bending are that the upper layers of material are stretched, while the lower layers are compressed[8]; a neutral layer in between is neither stretched nor compressed and y is measured with respect to this particular layer.

Geometrically, the effects of the bending can be visualised as in Figure 12.7. The neutral filament, which undergoes no change in length, is the curve AB; filaments above this are stretched, while filaments below are compressed. So filament CD stretches to CE and filament FG compresses to FH. The original right-hand end of the bar, DG, has been rotated through an angle θ to EH.

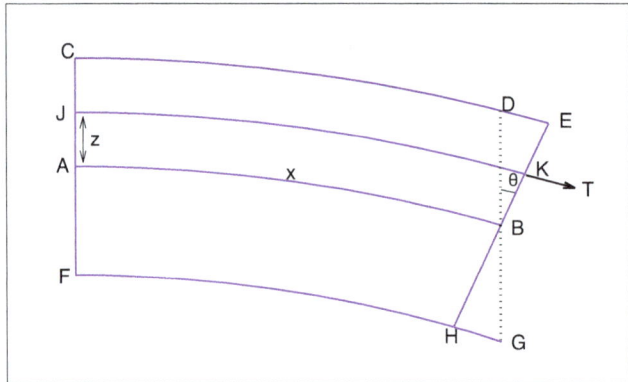

Figure 12.7: Elements of a bended beam

Consider just the stretched filament JK, a distance z above the neutral line. The extension — the amount it has stretched — is (approximately) $z\theta$, for small θ. The relative elongation, or *strain*, is then given by:

$$\text{strain} = \frac{z\theta}{x}.$$

---

[8]Think of bending a rectangular pencil eraser.

12.2. BEAM EQUATIONS

Suppose the force, or tension, acting at K is denoted by T. This is the force stretching the filament JK. Suppose also that the cross-sectional area of this filament is ΔA (a unitary quantity, like dA); then the *stress* (force per unit area) is:

$$\text{stress} = \frac{T}{\Delta A}.$$

Using the relation 'stress = $E\times$ strain', with E being Young's modulus[9], we must have:

$$\frac{T}{\Delta A} = E\frac{z\theta}{x}, \text{ which is equivalent to } T = \frac{Ez\theta\,\Delta A}{x}.$$

Now bring in the radius of curvature ρ of the filament. Since DG in Figure 12.7 is parallel with CF, it must be the case that EH makes the angle θ with CF; so extending both of these lines southwards until they meet will result in a point, distance ρ from both A and B. Thus,

$$x = \rho\theta,$$

resulting in

$$T = \frac{Ez\,\Delta A}{\rho}.$$

The clockwise bending moment M over the section at B is produced by the internal elastic forces, acting on the portion $CEHF$ by the rest of the beam. So taking moments about the horizontal line through B, we get:

$$M = \sum_z Tz, \text{ (summing over } z\text{)},$$
$$= \sum_z \frac{Ez^2\,\Delta A}{\rho},$$
$$= \frac{EI}{\rho},$$

where the quantity $I = \sum_z z^2\,\Delta A$ stands for the second moment of area of the cross-section at B about the horizontal line through B and perpendicular to the page. It is also called the geometrical moment of inertia of the cross-section, being proportional to the mechanical moment of inertia of a lamina[10] of the same shape as the beam cross-section [38].

We now make use of the approximation to the radius of curvature from Section 7.9, in which

$$\frac{1}{\rho} = \kappa \approx \frac{d^2y}{dx^2},$$

---

[9]Thomas Young, 1773 to 1829.
[10]A surface with negligible thickness; in the present case, imagine a very thin slice taken from the beam at x. So $I = \iint z^2 dA$ — a double integral over the cross-sectional area.

and so eventually end up with

$$M = -EI\frac{d^2y}{dx^2}, \qquad (12.12)$$

which relates the bending moment to the curvature of the beam. Note the negative sign that has been added here, which is a consequence of choosing a particular convention on the sign of M (see [42]).

As a final step (and it is the last step, at least in this section), bring in equation (12.11) to get:

$$EI\frac{d^4y}{dx^4} = f(x), \qquad (12.13)$$

assuming constant EI and — importantly — also assuming that the second derivative of M exists. This equation relates the actual downward bending y of the beam to the external downward force per unit length, f. If the latter is known, then y can be determined using repeated integration.

Should the reader be interested, a couple of example shapes of bending beams under different continuous and discrete loads are worked out in Appendix K.

Chapter 13

WORKING OUT AREAS

A further discussion of planar areas[1] does not sound particularly exciting, nor well-chosen to illustrate what maths can do, but there are some surprises. Working out the area of a rectangle is straightforward: it is just the product of the lengths of the two sides. And the area of a circle is πr^2, where r is the radius. More complicated or irregular shapes can, in principle, be determined by subdivision: divide into smaller triangles or rectangles, and add up the contributions. This can get rather tedious, however, and the accuracy of the result will depend on how well the subdivision is carried out.

The sections below illustrate two of the other, more unusual, methods that can be applied.

13.1 Tracing the Boundary

It may seem strange to be able to compute the area of some more or less complex shape simply by ambling around the perimeter, but this is exactly what the *planimeter* will do. An invention from Victorian times, the device is intended to be used in conjunction with a map: one moves a pointer completely around the mapped boundary (of a field, for example), reading the area from a dial at the end of the tour. But what is the calculation based on? In mathematical terms, the area A is given by [6]:

$$A = \frac{1}{2} \oint_C (x\,dy - y\,dx), \tag{13.1}$$

where the $\oint$ symbol stands for *contour integral*. The embedded circle and subscript C denote 'contour', or boundary, by convention traced in an anti-clockwise direction. The x and y refer to suitable Cartesian coordinates in the plane. It is assumed that the contour is 'simple', meaning primarily that it has no crossings[2].

---
[1] In the framework of Euclidean geometry; think of a flat sheet of paper.
[2] And the limiting case of the Koch curve in Chapter 26 is a further exception.

Equation 13.1 is a specific instantiation of Green's theorem[3] in the plane, which has more general applicability; Appendix G goes into a bit more detail.

Apply equation (13.1) now to the area in Figure 7.4 (see Section 7.6 above), using the contour shown below in Figure 13.1.

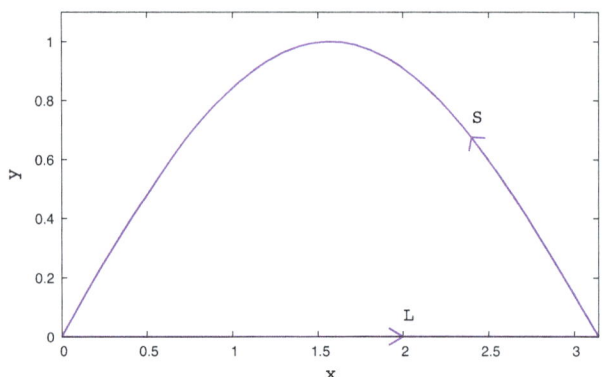

Figure 13.1: Contour for the sine area

Commencing at the origin at bottom-left, the contour C consists of a linear section along the bottom and call this L, followed by the curved bit around the top and call this S. So, splitting C into its two parts,

$$A = \frac{1}{2}\left[\int_L (x\,dy - y\,dx) + \int_S (x\,dy - y\,dx)\right],$$
$$= \frac{1}{2}\left[\int_L \left(x\frac{dy}{dx} - y\right)dx + \int_S \left(x\frac{dy}{dx} - y\right)dx\right],$$

integrating with respect to x in the second line. Throughout the contour section L, $y = 0$ and $dy/dx = 0$, so this contribution to the integral is zero. For contour S, it is known that $y = \sin x$ and so $dy/dx = \cos x$, giving

$$A = \frac{1}{2}\int_{x=\pi}^{0} (x\cos x - \sin x)\,dx.$$

Since the contour is traced anti-clockwise, the integral limits are from π to zero. It is not difficult to show that

$$\frac{d}{dx}(x\sin x + 2\cos x) = x\cos x - \sin x \quad \text{(see Section 7.10)},$$

so

$$A = \frac{1}{2}\Big[x\sin x + 2\cos x\Big]_\pi^0 = 2, \text{ as expected.}$$

---
[3] George Green, 1791 to 1841.

13.1. TRACING THE BOUNDARY

For digital computer applications, a more general instantiation of equation (13.1) can be developed when the boundary is either formed from straight line segments, or can be so approximated (any continuous curve can be approximated to the desired accuracy by making the segments short enough). So suppose that there are N such segments, enumerated using the index i, with i ranging from 1 to N inclusive, and with the last point coincident with the first (to create a closed boundary). Assume that line segment i joins the points $\{a_i, b_i\}$ to $\{a_{i+1}, b_{i+1}\}$. Then the area A may be written as a sum of integrals over the line segments, thus:

$$A = \frac{1}{2} \sum_{i=1}^{N-1} \int_i (x\,dy - y\,dx).$$

To aid the calculation of the integral over each segment, introduce the parameter s measured along that line, so that from point $\{a_i, b_i\}$ to $\{a_{i+1}, b_{i+1}\}$ we have:

$$x = x_i + \alpha_i s,$$
$$y = y_i + \beta_i s,$$

where s is in the range zero to L_i (the length of that line segment) and α_i, β_i (Greek 'beta') are parameters derivable from

$$\alpha_i = \frac{x_{i+1} - x_i}{L_i}, \quad \beta_i = \frac{y_{i+1} - y_i}{L_i}.$$

Concentrate now on the integral over line segment i, integrating with respect to s. So, for line segment i,

$$\begin{aligned} x\,dy - y\,dx &= \left(x\frac{dy}{ds} - y\frac{dx}{ds}\right) ds, \\ &= \left[(x_i + \alpha_i s)\beta_i - (y_i + \beta_i s)\alpha_i\right] ds, \\ &= (\beta_i x_i - \alpha_i y_i)\,ds. \end{aligned}$$

The integral over line segment i then becomes:

$$\begin{aligned} \int_i (x\,dy - y\,dx) &= \int_{s=0}^{L_i} (\beta_i x_i - \alpha_i y_i)\,ds, \\ &= L_i(\beta_i x_i - \alpha_i y_i), \end{aligned}$$

since x_i, y_i, α_i and β_i are all constant for this segment. Putting the bits together will give the area as a whole:

$$A = \frac{1}{2} \sum_{i=1}^{N-1} L_i(\beta_i x_i - \alpha_i y_i). \tag{13.2}$$

Given a digital file of two-dimensional points forming the boundary, this equation will provide the area. If, for example, the boundary of a square is defined by the five points $\{0,0\}, \{2,0\}, \{2,2\}, \{0,2\}$ and $\{0,0\}$, the area is found to be 4, as expected.

For a more interesting example, see Figure 13.2, which represents the boundary of a (hypothetical) piece of woodland. The units are metres, the curved sections are approximated by short linear steps and the arrow indicates the direction of travel.

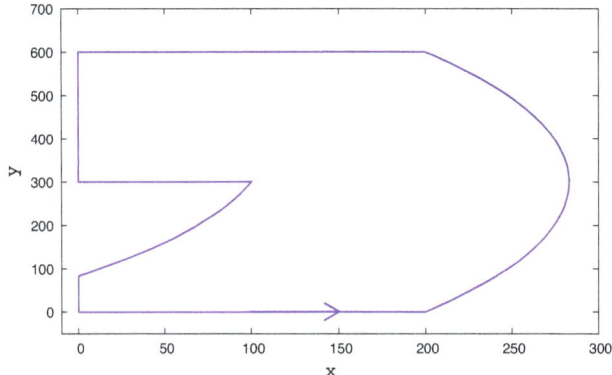

Figure 13.2: Hypothetical woodland boundary

The area of this works out to be 140737 square metres, or about 34.77 acres.

Quite how the planimeter implemented equation (13.1) in mechanical form is an interesting tale in itself, but not relevant to the discussion here. What is interesting is that equation (13.1) encapsulates the property of a two-dimensional area entirely in terms of its boundary, a finding that resurfaces in a related context in Section 17. There are also echoes here of the prediction that the information content of a black hole must reside entirely on its surface [43].

13.2 Monte Carlo Method

Here, it is shown how to estimate the area using random numbers. How can randomness help, we may ask, in this or anything else? In fact, mathematicians have found numerous ways of using randomness to advantage, this being one (and perhaps the simplest to appreciate).

Go back to the shaded area shown in Figure 7.4, in Section 7.6. As already mentioned, this represents the area under a half-cycle sine curve, from zero to π in radians, equivalent to zero to $180°$. Ask now what would happen if large numbers of uniformly distributed points are generated within the outer square area, in the region $0 \leq x \leq \pi$ and $0 \leq y \leq 1$. It is expected that a certain proportion of these points will fall within the purple shaded area, and that this numeric proportion will reflect the ratio of the purple area to the area of the outer square. 'Simples', as the meerkats might say.

It is necessary to determine whether any particular point falls within the purple region or not; a general method for doing this is examined later in Chap-

13.2. MONTE CARLO METHOD

ter 17.1, but the regular shape of the curve boundary makes it particularly simple here.

To illustrate the results, generate 4,000 uniformly-distributed random samples from within the outer area. The first 600 are shown in Figure 13.3, colour-coded depending on whether they were within the required area or not.

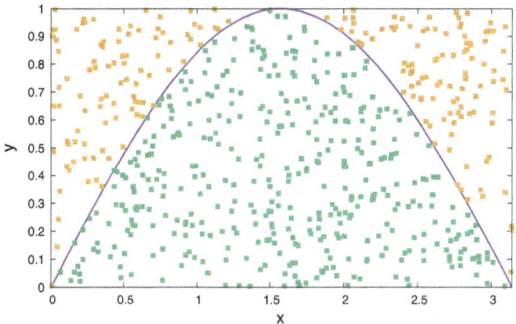

Figure 13.3: Example area with 600 random points

We can also see how the area calculation converges as the total number of random samples increases, giving the results shown in Figure 13.4. For reference, the exact area under the curve in Figure 7.4 is known to be 2 (Section 7.6)

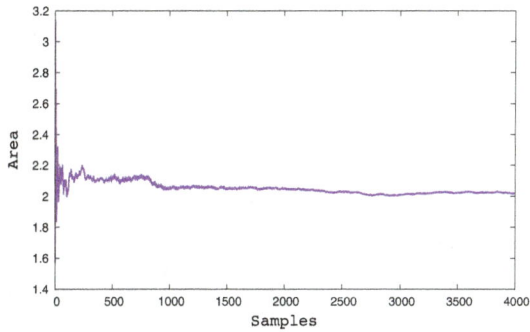

Figure 13.4: Convergence of the area estimate with sample size

As mentioned previously, the uncertainty of the estimate diminishes as the inverse square root of the number of samples (see Section 4.2), so this Monte Carlo method requires some patience.

This approach has wider application than in just the two dimensions used here; [32] applies it to a complex three-dimensional shape, where it might be difficult to apply more classical methods.

Chapter 14

VECTORS (AND RELATIVES)

J worked for a short while assisting in the development of software for one of the early medical devices for the treatment of deep brain cancers. The aim of the device was to irradiate the tumour using a pencil beam of X-rays, the narrow beam minimising damage to the surrounding brain tissue. However, using a single beam in a single direction over time will give essentially the same dose to all of the tissue along that beam, healthy as well as cancerous cells. So the idea was to move the X-ray source around the head, aiming from multiple different points but with all of the beams ending up crossing at the tumour site. Maximum dose for the tumour and minimum collateral damage.

It was natural, then, to model each beam as a *vector*. In its simplest form, this is just a group of numbers, such as the three-dimensional construct

$$\mathbf{s} = \{a,\ b,\ c\}\,.$$

The use of a bold font here denotes 'vector'; some notations use a little arrow over the top, like $\vec{s}$, but this is a bit fussier and I prefer to use what is familiar. In the present context, each of the numbers here can stand for one of the three spatial coordinates — say, in the x, y and z directions relative to some reference axis set. With the above X-ray device, such axes can be conveniently referenced to a rigid frame surrounding the patient's head.

Vectors of the same dimensionality can be added and subtracted, component-wise, as might be expected. So if $\mathbf{y} = \{d,\ e,\ f\}$, say, then

$$\mathbf{x} = \mathbf{s} + \mathbf{y} = \{a+d,\ b+e,\ c+f\}\,.$$

This process is illustrated in Figure 14.1, now using two spatial dimensions (consistent with the paper), and where

$$\mathbf{s} = \{2,\ 1\},$$
$$\mathbf{y} = \{1,\ 2\},$$
so $\mathbf{x} = \{3,\ 3\}\,.$

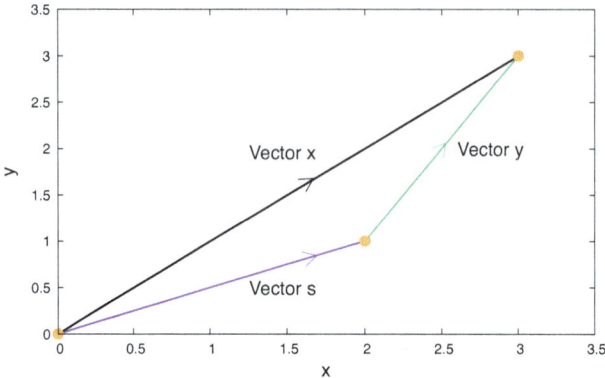

Figure 14.1: Illustration of vector addition

The point {3, 3} can be reached either by the direct line coloured blue, or via the more indirect route of purple followed by green. An instance of this is where the purple line represents where an aircraft is currently heading, while the green line stands for side-wind. The *actual* aircraft path over the ground will be given by the blue line.

Vector multiplication is a bit more complicated, and we'll come to that in a moment.

A vector may also be a *unit* vector, meaning of total length one. Using Pythagoras' theorem (again) and sticking with two spatial dimensions, such a vector $\hat{\mathbf{n}}$, with

$$\hat{\mathbf{n}} = \{p, q\},$$

would have $\sqrt{p^2 + q^2} = 1$. The little hat on top of the **n** here denotes unit length.

Now define

$$\mathbf{y} = \lambda \hat{\mathbf{n}},$$

thereby splitting the vector **y** into a positive length λ (Greek 'lambda') and a direction $\hat{\mathbf{n}}$. This subdivision is useful where we are not sure of the distance but we do know the direction.

Putting the various bits together, any three-dimensional location can be defined in terms of some reference point **s**, plus a direction given by $\hat{\mathbf{n}}$ and the distance λ. That is,

$$\mathbf{x} = \mathbf{s} + \lambda \hat{\mathbf{n}}.$$

This construct provides a suitable model for a pencil beam X-ray, since the location of the source **s** is known, as is the direction of transmission $\hat{\mathbf{n}}$. Any point along that beam can be defined simply by varying the parameter λ.

Now look at it in reverse: assume the tumour location **x** and the source of X-rays **s** are both known. So where do we point the beam if it is to pass through the tumour? Clearly,

$$\lambda \hat{\mathbf{n}} = \mathbf{x} - \mathbf{s}.$$

Since the direction is known to be a unit vector, the distance λ is derivable from the length of the vector $\mathbf{x} - \mathbf{s}$. This is written:

$$\lambda = |\mathbf{x} - \mathbf{s}|,$$

the vertical bars denoting a vector *norm*, or magnitude. So the direction will then be given by

$$\hat{\mathbf{n}} = \frac{\mathbf{x} - \mathbf{s}}{\lambda},$$

which means divide each component of vector $(\mathbf{x} - \mathbf{s})$ by λ.

A further vector operation will be needed: the *scalar product* of two vectors. This is just the sum of the products of the individual components. So if vector $\mathbf{x} = \{x_1, x_2\}$ and vector $\mathbf{y} = \{y_1, y_2\}$ (staying in two dimensions for the time being), then the scalar product is given by:

$$\mathbf{x} \cdot \mathbf{y} = x_1 y_1 + x_2 y_2.$$

From this, it is evident that

$$\mathbf{x} \cdot \mathbf{x} = x_1^2 + x_2^2 = |\mathbf{x}|^2,$$

in both cases with obvious generalisations to three (or, indeed, more) components.

We also have

$$\mathbf{x} \cdot \mathbf{y} = |\mathbf{x}||\mathbf{y}| \cos \theta,$$

where θ is the angle between the two vectors [44]. This is valid in two or three dimensions [6], and it may be observed that if the vectors are perpendicular (so that $\theta = \pi/2$), then their scalar product is zero.

Suppose now that the components of $\mathbf{x}$ are functions of some parameter t, say, and differentiate $\mathbf{x} \cdot \mathbf{x}$ with respect to t:

$$\frac{d}{dt}(\mathbf{x} \cdot \mathbf{x}) = \frac{d}{dt}\left(x_1^2 + x_2^2\right),$$

$$= 2x_1 \frac{dx_1}{dt} + 2x_2 \frac{dx_2}{dt},$$

$$= 2\mathbf{x} \cdot \frac{d\mathbf{x}}{dt}.$$

One more multiplicative vector operation that will be needed later on in the book is the *vector product* or *cross product*. Defined in three dimensional space, this is as follows:

$$\mathbf{x} \wedge \mathbf{y} = |\mathbf{x}||\mathbf{y}| \sin \theta; \text{ (this is sometimes written } \mathbf{x} \times \mathbf{y}\text{)},$$

where θ is again the angle between $\mathbf{x}$ and $\mathbf{y}$; and $\mathbf{x} \wedge \mathbf{y}$ is perpendicular to both $\mathbf{x}$ and $\mathbf{y}$.

In component terms, if $\mathbf{x} = (x_1, x_2, x_3)$ and $\mathbf{y} = (y_1, y_2, y_3)$, then

$$\mathbf{x} \wedge \mathbf{y} = (x_2 y_3 - x_3 y_2,\ x_3 y_1 - x_1 y_3,\ x_1 y_2 - x_2 y_1).$$

With the above fundamentals under our belt, it is now possible to look at one of the required functions of the X-ray machine: given two bearings emanating from two different source locations but pointing more or less to the same place, how far apart are they at closest approach? This is most easily visualised in three dimensions by taking two straight bamboo stakes (or drinking straws, or similar), one in each hand, and bringing the stick centres nearly but not quite touching[1].

The 'brute force and ignorance' approach (as would have been disparagingly referred to by one of my university lecturers) is one way of finding the point of closest approach: start two counters at each source location and progress each one in small steps along in its particular direction, stopping when the separation is a minimum.

A more elegant (and more accurate) solution is obtained using calculus. Suppose the two source locations are labelled $\mathbf{s}_1$ and $\mathbf{s}_2$, with their associated directions being $\hat{\mathbf{n}}_1$ and $\hat{\mathbf{n}}_2$. Then two arbitrary points along each of the two directions will be given by the vectors:

$$\mathbf{x}_1 = \mathbf{s}_1 + \lambda_1\, \hat{\mathbf{n}}_1, \tag{14.1}$$
$$\mathbf{x}_2 = \mathbf{s}_2 + \lambda_2\, \hat{\mathbf{n}}_2, \tag{14.2}$$

and the squared distance between these two points will be:

$$J = (\mathbf{x}_2 - \mathbf{x}_1) \cdot (\mathbf{x}_2 - \mathbf{x}_1),$$

using the scalar product. The reason for choosing a squared distance will become apparent below.

The quantity J can be regarded as a function of the two independent parameters λ_1 and λ_2, since all of the other quantities are known. It can therefore be regarded as a two-dimensional surface — think of a rubber sheet with curvature in two directions. So to find the *minimum* value of J (or, more generally, the extrema), differentiate with respect to both parameters to get the function slopes and set both slopes to zero:

$$\frac{\partial J}{\partial \lambda_1} = 0 \text{ and } \frac{\partial J}{\partial \lambda_2} = 0,$$

which imply that

$$(\mathbf{x}_2 - \mathbf{x}_1) \cdot \frac{\partial}{\partial \lambda_1}(\mathbf{x}_2 - \mathbf{x}_1) = 0,$$
$$(\mathbf{x}_2 - \mathbf{x}_1) \cdot \frac{\partial}{\partial \lambda_2}(\mathbf{x}_2 - \mathbf{x}_1) = 0.$$

---

[1] These are known as skew bearings: non-parallel and not in the same plane.

Since $\mathbf{x}_1$ is a function only of λ_1 (from equation (14.1)), and $\mathbf{x}_2$ is a function only of λ_2, we find that:

$$\frac{\partial}{\partial \lambda_1}(\mathbf{x}_2 - \mathbf{x}_1) = -\hat{\mathbf{n}}_1,$$

$$\frac{\partial}{\partial \lambda_2}(\mathbf{x}_2 - \mathbf{x}_1) = \hat{\mathbf{n}}_2,$$

so the equations for the minimum become, after some rearrangement:

$$\hat{\mathbf{n}}_1 \cdot (\mathbf{x}_2 - \mathbf{x}_1) = 0, \tag{14.3}$$
$$\hat{\mathbf{n}}_2 \cdot (\mathbf{x}_2 - \mathbf{x}_1) = 0. \tag{14.4}$$

Now substitute equations (14.1) and (14.2) into $\mathbf{x}_2 - \mathbf{x}_1$, to get:

$$\mathbf{x}_2 - \mathbf{x}_1 = (\mathbf{s}_2 - \mathbf{s}_1) + \lambda_2\,\hat{\mathbf{n}}_2 - \lambda_1\,\hat{\mathbf{n}}_1,$$

and chuck this into equations (14.3) and (14.4):

$$\hat{\mathbf{n}}_1 \cdot (\mathbf{s}_2 - \mathbf{s}_1) + \lambda_2\,\hat{\mathbf{n}}_1 \cdot \hat{\mathbf{n}}_2 - \lambda_1 = 0,$$
$$\hat{\mathbf{n}}_2 \cdot (\mathbf{s}_2 - \mathbf{s}_1) + \lambda_2 - \lambda_1\,\hat{\mathbf{n}}_1 \cdot \hat{\mathbf{n}}_2 = 0,$$

using the fact that $\hat{\mathbf{n}}_1$ and $\hat{\mathbf{n}}_2$ are unit vectors.

Rearrange these two equations to put the unknown quantities on the left-hand-side and the known stuff on the right:

$$\lambda_1 - \lambda_2\,\hat{\mathbf{n}}_1 \cdot \hat{\mathbf{n}}_2 = \hat{\mathbf{n}}_1 \cdot (\mathbf{s}_2 - \mathbf{s}_1),$$
$$\lambda_1\,\hat{\mathbf{n}}_1 \cdot \hat{\mathbf{n}}_2 - \lambda_2 = \hat{\mathbf{n}}_2 \cdot (\mathbf{s}_2 - \mathbf{s}_1).$$

These form two linear equations in λ_1 and λ_2, and a small amount of algebra will give the required solution. It is more instructive, though, to combine both equations into a single one involving a *matrix*[2] and two vectors:

$$\begin{pmatrix} 1 & -\hat{\mathbf{n}}_1 \cdot \hat{\mathbf{n}}_2 \\ \hat{\mathbf{n}}_1 \cdot \hat{\mathbf{n}}_2 & -1 \end{pmatrix} \begin{pmatrix} \lambda_1 \\ \lambda_2 \end{pmatrix} = \begin{pmatrix} \hat{\mathbf{n}}_1 \cdot (\mathbf{s}_2 - \mathbf{s}_1) \\ \hat{\mathbf{n}}_2 \cdot (\mathbf{s}_2 - \mathbf{s}_1) \end{pmatrix}.$$

This can be written as:

$$M\mathbf{L} = \mathbf{R},$$

where

$$\mathbf{L} = \begin{pmatrix} \lambda_1 \\ \lambda_2 \end{pmatrix} \text{ and } \mathbf{R} = \begin{pmatrix} \hat{\mathbf{n}}_1 \cdot (\mathbf{s}_2 - \mathbf{s}_1) \\ \hat{\mathbf{n}}_2 \cdot (\mathbf{s}_2 - \mathbf{s}_1) \end{pmatrix},$$

both of which are vectors written in column form, rather than with components in a row. The remaining construct M with two rows and two columns is called

---

[2] Nothing to do with the movie of the same name, but one of the afore-mentioned vector relatives.

a matrix, in this case of square form, although different numbers of rows and columns are possible.

Multiplying a matrix by a vector, or multiplying two matrices together, is a bit more complicated than for pure numbers. It is most easily explained using index notation, in which two indices (or counters) are used for M and one index for **L**:

$$M\mathbf{L} \text{ is equivalent to } \sum_j M_{ij} L_j,$$

where i labels the rows and j the columns. To keep track of the various combinations, I find it convenient to use one finger on the left hand to move along the rows and one finger on the right hand to move down the columns, stepping each in unison.

The index notation also enables a more or less direct translation of matrix-vector arithmetic into computer code.

A square matrix can have an inverse, written as M^{-1}, so that in the present case

$$MM^{-1} = M^{-1}M = \begin{pmatrix} 1 & 0 \\ 0 & 1 \end{pmatrix},$$

the final quantity being the unit matrix.

On this basis,

$$\mathbf{L} = M^{-1}\mathbf{R}, \tag{14.5}$$

thus giving the solution via matrices.

But what is the inverse M^{-1}? For a general two-dimensional square matrix, A, say, with

$$A = \begin{pmatrix} a & b \\ c & d \end{pmatrix}, \text{ the inverse is: } A^{-1} = \frac{1}{ad - bc}\begin{pmatrix} d & -b \\ -c & a \end{pmatrix},$$

where the multiplicative factor $1/(ad - bc)$ multiplies each element. Inverting higher-dimensional matrices is rather more complicated and generally needs to be carried out numerically.

In the present case,

$$M = \begin{pmatrix} 1 & -\hat{\mathbf{n}}_1 \cdot \hat{\mathbf{n}}_2 \\ \hat{\mathbf{n}}_1 \cdot \hat{\mathbf{n}}_2 & -1 \end{pmatrix}, \text{ so } M^{-1} = \frac{1}{\left[-1 + (\hat{\mathbf{n}}_1 \cdot \hat{\mathbf{n}}_2)^2\right]}\begin{pmatrix} -1 & \hat{\mathbf{n}}_1 \cdot \hat{\mathbf{n}}_2 \\ -\hat{\mathbf{n}}_1 \cdot \hat{\mathbf{n}}_2 & 1 \end{pmatrix}$$

The denominator quantity, $-1 + (\hat{\mathbf{n}}_1 \cdot \hat{\mathbf{n}}_2)^2$ — called the *determinant* — is worth a second look. If the two direction vectors $\hat{\mathbf{n}}_1$ and $\hat{\mathbf{n}}_2$ happen to be parallel, then the determinant will be zero and no solution exists — there is no minimum. More generally, the set of linear equations $M\mathbf{L} = \mathbf{R}$ will only have a solution if M is non-singular — if it has a finite inverse.

Given M^{-1} and $\mathbf{R}$, the vector $\mathbf{L}$ is derivable from equation (14.5) and the two components of $\mathbf{L}$ are λ_1 and λ_2.

14.1 Normal Vector to a Surface

Suppose a surface embedded in a three-dimensional Cartesian space is defined by the continuous function $\phi(x,y,z) = 0$. By way of example, the surface of an ellipsoid is given by

$$\frac{x^2}{a^2} + \frac{y^2}{b^2} + \frac{z^2}{c^2} = 1, \text{ in which case } \phi = \frac{x^2}{a^2} + \frac{y^2}{b^2} + \frac{z^2}{c^2} - 1,$$

and where a, b and c are constants defining the semi-axes of the volume.

Now define some arbitrary curve C lying entirely on the surface and measure path length s along it relative to some convenient starting point. This curve can be specified parametrically by $x(s)$, $y(s)$ and $z(s)$ with s as parameter, so that

$$\phi(x(s), y(s), z(s)) = 0,$$

since C lies on the surface.

Differentiate ϕ with respect to s, thus giving the rate of change of ϕ along the path:

$$\frac{d\phi}{ds} = \frac{\partial \phi}{\partial x}\frac{dx}{ds} + \frac{\partial \phi}{\partial y}\frac{dy}{ds} + \frac{\partial \phi}{\partial z}\frac{dz}{ds} = 0.$$

This can also be written as the vector scalar product:

$$\nabla \phi \cdot \left(\frac{dx}{ds}, \frac{dy}{ds}, \frac{dz}{ds}\right) = 0, \qquad (14.6)$$

where

$$\nabla \phi = \left(\frac{\partial \phi}{\partial x}, \frac{\partial \phi}{\partial y}, \frac{\partial \phi}{\partial z}\right).$$

The vector $(dx/ds, dy/ds, dz/ds)$ lies along the curve, and indeed is tangential to it at each point, so it must also be tangential to the surface. This tangential property can be appreciated by considering a point P lying on C with Cartesian coordinates $(x(s), y(s), z(s))$, as well as a nearby point Q, also on C and with coordinates $(x(s+\delta s), y(s+\delta s), z(s+\delta s))$, the quantity δs specifying the (infinitesimal) distance along the curve from P to Q.

Therefore, expanding $x(s+\delta s)$ in a Taylor series, the point Q can be approximated:

$$Q: \left(x(s) + \delta s \frac{dx}{ds} + O(\delta s^2), y(s) + \delta s \frac{dy}{ds} + O(\delta s^2), z(s) + \delta s \frac{dz}{ds} + O(\delta s^2)\right),$$

$$= (x(s), y(s), z(s)) + \delta s \left(\frac{dx}{ds}, \frac{dy}{ds}, \frac{dz}{ds}\right) + O(\delta s^2)$$

So vector $(dx/ds, dy/ds, dz/ds)$ must point along the curve and be tangential to the curve at P. In which case the quantity $\nabla \phi$ must be normal to the surface — i.e., it is a vector pointing perpendicularly away from the surface — since if the scalar product of two vectors is zero, the vectors are normal to each other[3].

The same logic applies if we want the normal vector to a two-dimensional curve, for example the ellipse

$$\phi(x,y) = \frac{x^2}{a^2} + \frac{y^2}{b^2} - 1 = 0, \tag{14.7}$$

in which a and b are constants. The normal vector at a point $P : (x, y)$ will be given as

$$\nabla \phi = \left(\frac{\partial \phi}{\partial x}, \frac{\partial \phi}{\partial x}\right) = 2\left(\frac{x}{a^2}, \frac{y}{b^2}\right), \tag{14.8}$$

and this is illustrated in Figure 14.2 pointing out from the curve[4]. Its back-projection to the x-axis is added to show that such normal vectors need not pass through the origin.

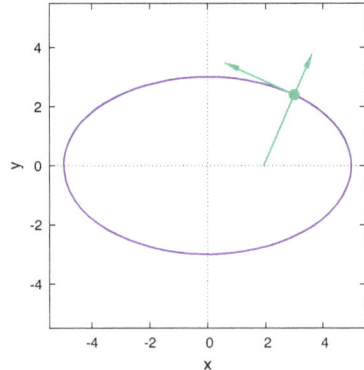

Figure 14.2: Illustration of normal and tangent vectors to an ellipse

The tangent vector $\hat{\mathbf{t}}$ is also illustrated, this being given by the definition:

$$\hat{\mathbf{t}} = \left(\frac{dx}{ds}, \frac{dy}{ds}\right).$$

What is the path length s in this case, so that we may derive the derivatives of x and y with respect to it? From an inspection of equation (14.7), a parametrisation of x and y can be obtained using the angle θ relative to the origin:

$$x = a \cos\theta, \ y = b \sin\theta,$$

but θ is not the same as s, θ being dimensionless whereas s has units of length.

---

[3] Ignoring the trivial case where one or other vector is zero.
[4] Neither vector shown is of unit length.

14.1. NORMAL VECTOR TO A SURFACE

The path length s can be obtained straightforwardly from the infinitesimal Pythagorean formula:

$$ds^2 = dx^2 + dy^2 = \left[\left(\frac{dx}{d\theta}\right)^2 + \left(\frac{dy}{d\theta}\right)^2\right] d\theta^2,$$
$$= [a^2 \sin^2 \theta + b^2 \cos^2 \theta] \, d\theta^2. \tag{14.9}$$

from the definitions of x, y in terms of θ.

Solving for s can then be achieved by integrating to get s as a function of θ, via

$$s = \int \sqrt{a^2 \sin^2 \theta + b^2 \cos^2 \theta} \, d\theta,$$

although this step does require engaging with elliptic functions [36]. Fortunately, for present purposes, such complexity is not needed, since it is only dx/ds and dy/ds that are required, and these are given simply in terms of θ by:

$$\frac{dx}{ds} = \frac{dx}{d\theta}\frac{d\theta}{ds}, \quad \frac{dy}{ds} = \frac{dy}{d\theta}\frac{d\theta}{ds}.$$

That is, using equation (14.9),

$$\frac{dx}{ds} = \frac{-a \sin \theta}{\sqrt{a^2 \sin^2 \theta + b^2 \cos^2 \theta}}, \quad \frac{dy}{ds} = \frac{b \cos \theta}{\sqrt{a^2 \sin^2 \theta + b^2 \cos^2 \theta}}.$$

By using the definitions of x and y in terms of θ, these may also be written (after a bit of algebra) as:

$$\frac{dx}{ds} = \frac{-a^2 y}{\sqrt{a^4 y^2 + b^4 x^2}}, \quad \frac{dy}{ds} = \frac{b^2 x}{\sqrt{a^4 y^2 + b^4 x^2}}.$$

Therefore, the tangent vector in terms of x, y and associated constants is:

$$\hat{\mathbf{t}} = \frac{1}{\sqrt{a^4 y^2 + b^4 x^2}} \left(-a^2 y, b^2 x\right).$$

A quick check shows that this is perpendicular to the $\nabla \phi$ given in equation (14.8), and also that $\hat{\mathbf{t}}$ is a unit vector (which is not the case for $\nabla \phi$).

Chapter 15

CIRCUITS AND RESONANCE

THIS chapter takes a brief look at differential equations — the sort termed 'ordinary', in which the derivatives are with respect to one variable only (time in this case). We wish to analyse the response of the electrical circuit illustrated in Figure 15.1.

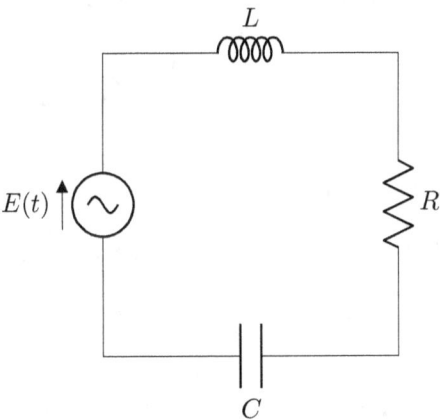

Figure 15.1: Simple electrical circuit

This shows a time-varying power source, $E(t)$, connected in series to an inductor L, a resistor R and a capacitor C. Suppose that the input electromotive force is given by $E(t) = E_0 \sin \omega t$; how, then, does the current I in the circuit behave over time?

The responses of each of the L, C and R components is given by the following relations [6]:

$$\text{Voltage drop across an inductor}: \quad E_L = L\frac{dI}{dt},$$

$$\text{Voltage drop across a capacitor}: \quad E_C = \frac{1}{C}\int I(t)\,dt,$$

$$\text{Voltage drop across a resistor}: \quad E_R = RI.$$

From Kirchhoff's voltage law[1], we have:

$$E(t) = E_L + E_C + E_R,$$

thus giving the integro-differential equation:

$$E_0 \sin\omega t = L\frac{dI}{dt} + \frac{1}{C}\int I(t)\,dt + RI.$$

This can be put into a more amenable form for solution by differentiating once with respect to time:

$$L\frac{d^2I}{dt^2} + R\frac{dI}{dt} + \frac{1}{C}I = \omega E_0 \cos\omega t, \tag{15.1}$$

after a bit of reordering. This is now a second-order ordinary differential equation with constant coefficients.

Before proceeding to a solution, a bit of background on what the various electrical components do may be helpful. A resistor does essentially what it says: it directly impedes the flow of current, typically becoming warm in the process. A capacitor stores energy, while an inductor provides impedance in a different way: it responds to the varying current by generating a magnetic field that in turn opposes the motion of electrons in the wire.

Now for the solution, which for convenience is generally divided into two parts: a *general solution* for which the right-hand-side is zero; and a *particular solution* taking specific account of the $\cos\omega t$ driving term.

General Solution

The general solution gives the innate response of the circuit in the absence of any forcing function, and can be found by solving the equation

$$L\frac{d^2I}{dt^2} + R\frac{dI}{dt} + \frac{1}{C}I = 0.$$

Recall from Chapter 7 that the derivative of $e^{\lambda x}$ is $\lambda e^{\lambda x}$, so suppose that $I = e^{\lambda t}$ for some parameter λ. Plugging this into the above equation then gives:

$$e^{\lambda t}\left(L\lambda^2 + R\lambda + \frac{1}{C}\right) = 0,$$

---
[1] Gustav Kirchhoff, 1824 to 1887.

here separating out the common factor $e^{\lambda t}$. Since the exponential function is only zero at $t=$ negative infinity, the practical solution is given by

$$L\lambda^2 + R\lambda + \frac{1}{C} = 0,$$

which implies two possible values of λ:

$$\lambda = \frac{1}{2L}\left\{-R \pm \sqrt{R^2 - 4L/C}\right\}.$$

That is,

$$\lambda_1 = \frac{1}{2L}\left\{-R + \sqrt{R^2 - 4L/C}\right\}, \tag{15.2}$$

$$\lambda_2 = \frac{1}{2L}\left\{-R - \sqrt{R^2 - 4L/C}\right\}. \tag{15.3}$$

For the interested reader, the method for solving general quadratic equations is briefly discussed in Appendix E.

Particular Solution

The particular solution copes with the specific form of the forcing function on the right-hand-side. Looking at the form of equation (15.1), suppose I can be written as a sum of two sinusoidal terms:

$$I = \alpha \cos \omega t + \beta \sin \omega t, \tag{15.4}$$

for some parameters α and β. Substitute this into equation (15.1), carry out the derivatives and collect terms:

$$\cos \omega t \left(-\alpha \omega^2 L + \beta \omega R + \frac{\alpha}{C}\right) + \sin \omega t \left(-\beta \omega^2 L - \alpha \omega R + \frac{\beta}{C}\right) \equiv \omega E_0 \cos \omega t.$$

The symbol $\equiv$ stands for 'identically equal', meaning in the present case that cosine terms on the left should match cosine terms on the right and similarly for the sine terms. So we get a pair of simultaneous algebraic equations in α and β

$$-\alpha \omega^2 L + \beta \omega R + \frac{\alpha}{C} = \omega E_0,$$

$$-\beta \omega^2 L - \alpha \omega R + \frac{\beta}{C} = 0.$$

Now borrow some of the matrix techniques from Chapter 14 and write these two equations in the form:

$$\begin{pmatrix} (\frac{1}{C} - \omega^2 L) & \omega R \\ -\omega R & (\frac{1}{C} - \omega^2 L) \end{pmatrix} \begin{pmatrix} \alpha \\ \beta \end{pmatrix} = \begin{pmatrix} \omega E_0 \\ 0 \end{pmatrix},$$

and inverting the matrix gives

$$\begin{pmatrix} \alpha \\ \beta \end{pmatrix} = \frac{1}{(\omega R)^2 + (\frac{1}{C} - \omega^2 L)^2} \begin{pmatrix} (\frac{1}{C} - \omega^2 L) & -\omega R \\ \omega R & (\frac{1}{C} - \omega^2 L) \end{pmatrix} \begin{pmatrix} \omega E_0 \\ 0 \end{pmatrix}.$$

Solving for α and β:

$$\alpha = \frac{\omega E_0 \left(\frac{1}{C} - \omega^2 L\right)}{\left[(\omega R)^2 + \left(\frac{1}{C} - \omega^2 L\right)^2\right]},$$

$$\beta = \frac{\omega^2 R E_0}{\left[(\omega R)^2 + \left(\frac{1}{C} - \omega^2 L\right)^2\right]}.$$

Complete Solution

Putting the bits together, the complete solution for the current I will be given by the sum of the general and particular solutions:

$$I = A e^{\lambda_1 t} + B e^{\lambda_2 t} + \frac{\omega E_0}{D} \left\{ \left(\frac{1}{C} - \omega^2 L\right) \cos \omega t + \omega R \sin \omega t \right\}, \quad (15.5)$$

where λ_1, λ_2 are defined above, A, B are constants, and

$$D = (\omega R)^2 + \left(\frac{1}{C} - \omega^2 L\right)^2 \quad (15.6)$$

has been substituted for later reference.

By way of example, suppose $R = 30$ ohms (the unit of resistance), $L = 0.2$ henrys (the unit of inductance) and $C = 0.002$ farad (the unit of capacitance). Set ω to be consistent with 60 Hz (60 cycles per second), giving $\omega = 2\pi \times 60$, with $E_0 = 5$ volts. Assume at the outset, $t = 0$, that the current in the circuit is zero and ditto its rate of change. Then pull the switch, so $E(t)$ acts immediately after $t = 0$. How does the current then behave with time?

Graphically, the response is shown in Figure 15.2. Apart from the first 0.03 seconds or so, the current settles down into a steady sinusoidal behaviour, governed essentially by the last two terms in equation (15.5).

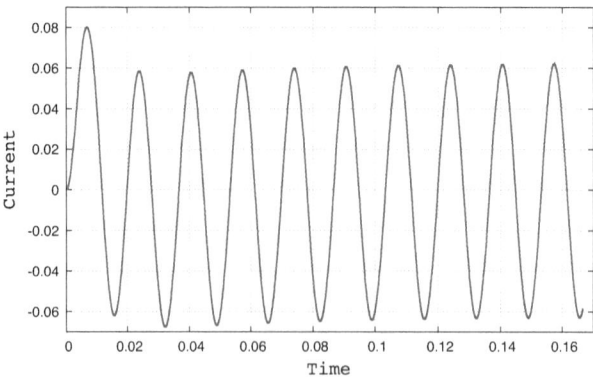

Figure 15.2: Current in electrical circuit

To understand this, it is not actually necessary to determine A and B from the initial conditions. The first two terms in equation (15.5) are of exponential form, which could either increase or decrease with time. To find out which, proceed on the assumption that R, L and C are all positive and that $R^2 > 4L/C$. Then, from equation (15.3), λ_2 must be negative. Also, since $4L/C < R^2$, the square root quantity in λ_1 must be less than R and so $\lambda_1 < 0$ as well.

So both of the exponential terms decay with time, leaving the longer-term behaviour of I governed by the purely sinusoidal terms in equation (15.5). In other words, apart from some transient behaviour just after the switch is pulled, the current oscillates with the same frequency ω as $E(t)$.

This is not the end of the story, though — more information can be pulled out of equation (15.5). Assume that we have gone far enough along in time to ignore the exponential terms, so that the current is governed by the following equation:

$$I \approx \frac{\omega E_0}{D} \left\{ \left(\frac{1}{C} - \omega^2 L\right) \cos \omega t + \omega R \sin \omega t \right\}.$$

By making use of the equation $\sin(\omega t + \psi) = \sin \omega t \cos \psi + \cos \omega t \sin \psi$ (from Section 7.3), the current may also be written in the form:

$$I \approx \frac{\omega E_0}{\sqrt{D}} \sin(\omega t + \psi), \qquad (15.7)$$

where the phase angle ψ is given by

$$\tan \psi = \frac{1}{\omega R}\left(\frac{1}{C} - \omega^2 L\right).$$

The merit of obtaining equation (15.7) (apart from the obvious simplification from its predecessor) is that the *amplitude* of the current response is immediately visible as $\omega E_0/\sqrt{D}$, since the sine function is limited in magnitude to ± 1. This suggests that varying the circuit parameters R, L and C can be used to alter the magnitude of the current — in effect, to tune the response.

Going back to the definition of D from equation (15.6), we may now ask what values of R, L and C will minimise D, since a small value of D will increase I (from equation (15.7)). It would be tempting to set $R = 0$, but this is unachievable in practice (except in the case of superconductivity, but see below in Section 15.1), so the best that can be done is to set one or other (or both) of L and C such that

$$LC = \frac{1}{\omega^2}.$$

This forms the basis for circuit tuning, in that L, say, can be varied to concentrate on one incoming frequency among the medley present in the antenna signal. An old-fashioned radio tuning dial would be doing essentially this.

15.1 Resonance

Go back now to the original circuit diagram in Figure 15.1 and ask what would happen if the resistance R was zero *and* $LC = 1/\omega^2$. From equations (15.6) and (15.5) it might be presumed that the current has to be infinite, since $D = 0$. But fortunately not so, since the particular solution defined above in equation (15.4) was calculated on the assumption that it would involve different functions from those in the general solution.

To see this, set $R = 0$ and $LC = 1/\omega^2$; the differential equation governing the circuit behaviour is now

$$\frac{d^2 I}{dt^2} + \omega^2 I = \frac{\omega E_0}{L} \cos \omega t, \tag{15.8}$$

(on the assumption that L is not zero). The general solution now involves $\sin \omega t$ and $\cos \omega t$, which overlap the $\cos \omega t$ on the right-hand-side, so something different for the particular solution is required; we cannot reuse equation (15.4). There is a general equation involving integrals for obtaining the particular solution [6], but a simpler approach in the present case is to try something of the form

$$I = \alpha t \cos \omega t + \beta t \sin \omega t.$$

Plugging this in to equation (15.8) and equating the terms results in

$$I = \frac{E_0 t}{2L} \sin \omega t.$$

In fact, due to the initial conditions, this is the entire solution. It is still sinusoidal in form, but with an amplitude that grows linearly with time, as can be appreciated from Figure 15.3.

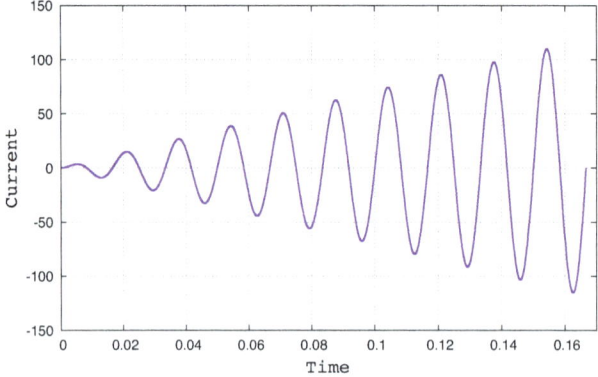

Figure 15.3: Current in electrical circuit with resonance

This sort of behaviour is termed *resonance*, where the forcing function has a frequency that matches (or nearly matches) the innate (or 'natural') frequency

15.1. RESONANCE

of vibration. Resonance can occur in mechanical systems as well; the reader may have noticed that some tall and thin industrial chimneys have 'fins' spiralling around part of the height. These act to disrupt periodic wind-induced vortex shedding, which otherwise can occur (depending on the wind speed) with a frequency similar to the natural oscillation frequency of the structure. If the response was anything like Figure 15.3, the chimney would not last long.

As to how one determines the natural frequencies of a structure, an approach (other than experimentation) is to give it a prod (mathematically speaking) and predict how it will respond. An example of this can be found in Section 25.4) below.

Chapter 16

FREQUENCY ANALYSIS

TAKE a brief look at Figure 16.1 and ask how the information contained in the curve here could most efficiently be transmitted over some data link — perhaps sent to a friend or colleague by email, for example.

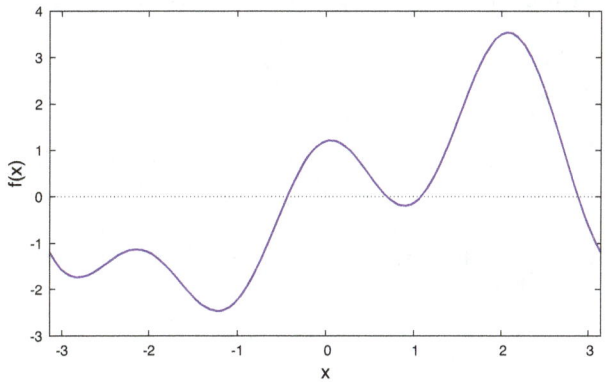

Figure 16.1: Example sinusoidal sequence

Sure, it is possible to construct a lengthy list of numbers, starting on the left and working towards the right in small steps, thereby 'drawing' the curve in numeric terms. But this uses a lot of data for what is actually a very concise expression, since the curve is specified exactly by

$$f(x) = 2\sin x - 0.7\sin 2x + 1.2\cos 3x, \qquad (16.1)$$

in the range $[-\pi, \pi]$. The quantity $\sin 2x$ stands for $\sin(2x)$ — double the frequency of $\sin x$ by itself — and so on up though higher harmonics.

Thus Figure 16.1 in x-space is transformed into Figure 16.2 in frequency space: just three numbers are needed.

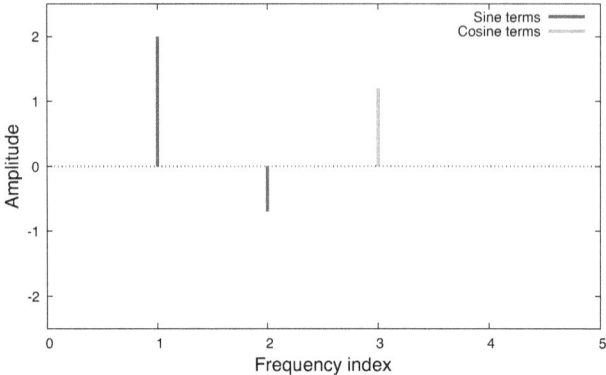

Figure 16.2: The same data mapped to frequency space

Transmitting this information is much more efficient (at least given an agreed protocol enabling the reconstruction of the curve at the other end).

In more general terms, it turns out that almost any periodic function can be expressed in terms of a trigonometric series such as equation (16.1). For a function $f(x)$ with period 2π, the general *Fourier series*[1] is written

$$f(x) = a_0 + \sum_{n=1}^{\infty} \left(a_n \cos nx + b_n \sin nx \right). \tag{16.2}$$

The a_n, b_n quantities are coefficients and n is an integer. The upper bound ∞ states that the series may theoretically progress without limit, although in practice it is necessary to terminate the terms at some convenient point.

Given some periodic function $f(x)$, the coefficients in the series are then derived using the following integrals [6]:

$$\begin{aligned}
a_0 &= \frac{1}{2\pi} \int_{-\pi}^{\pi} f(x)\,dx, \\
a_n &= \frac{1}{\pi} \int_{-\pi}^{\pi} f(x) \cos nx\, dx, \quad n \geq 1, \\
b_n &= \frac{1}{\pi} \int_{-\pi}^{\pi} f(x) \sin nx\, dx, \quad n \geq 1.
\end{aligned} \tag{16.3}$$

Apply this now to the discontinuous square-wave function, whereby

$$f(x) = \begin{cases} -1 & \text{for } -\pi < x < 0, \\ 1 & \text{for } 0 < x < \pi. \end{cases}$$

The results of summing equation (16.2) for various upper bounds $n \leq N$ is shown in Figure 16.3. The more terms that are summed, the better the functional representation becomes.

---

[1] Joseph Fourier, 1768 to 1830.

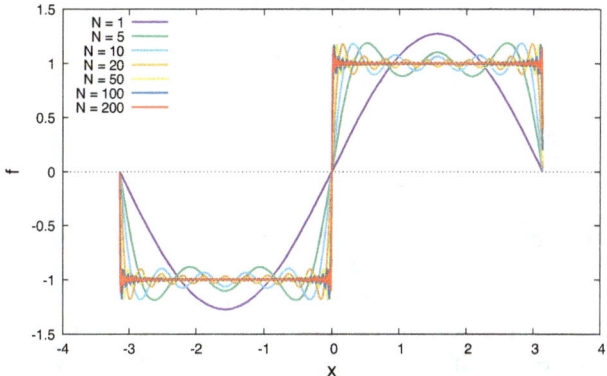

Figure 16.3: Fourier representation of the square wave

It turns out that all of the a_n are zero, while the b_n are given by:

$$b_n = \frac{2}{n\pi}(1 - \cos n\pi),$$

which are shown in the frequency plot of Figure 16.4, up to $n = 50$.

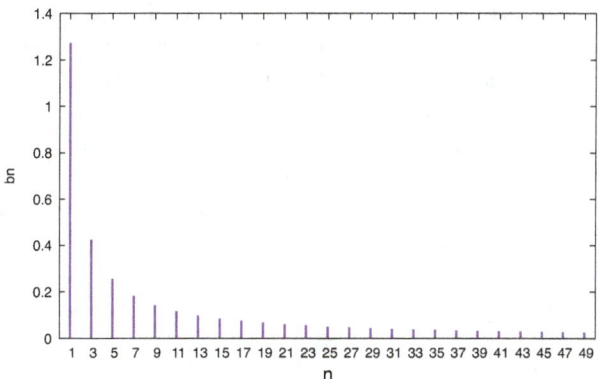

Figure 16.4: Fourier spectrum for the square wave

All of the even values of n generate $b_n = 0$, so only the odd values of n are relevant. So the continuous square wave can be represented by the discrete set of odd b_n values, to whatever accuracy is needed.

Equation (16.2) is applicable to periodic functions. For *non*-periodic functions, it is necessary to use Fourier integrals of the form [6]

$$f(x) = \int_{w=0}^{\infty} \left[A(w) \cos wx + B(w) \sin wx \right] dw,$$

with some restrictions on the form of f. The coefficients in this case are given by:

$$A(w) = \frac{1}{\pi} \int_{-\infty}^{\infty} f(s) \cos ws \, ds,$$

$$B(w) = \frac{1}{\pi} \int_{-\infty}^{\infty} f(s) \sin ws \, ds.$$

Fourier integrals form part of a set of integral transform techniques, and the topic is explored further in Chapter 20.

Such spectral analysis as was examined above is not just used for efficient data transmission, but has much wider applicability. For instance, some years ago, I was asked to examine sound recordings made by heavy machinery. The machinery operatives[2] had, after long acclimatisation, noticed that the sounds made by these machines changed discernably prior to failure, so the intent was to see if incipient failure could be automatically detected. The first step in this project was to acquire sound recordings made by these machines in both normal operation and when tending toward failure, in the operatives' opinion. Frequency analysis of the above form was the first step, to see if the spectral patterns in frequency space showed differences across the two machine categories. Unfortunately, the project was never initiated due to lack of funding, but there is every expectation that such spectral differences would have been seen.

Real-life examples of spectral analysis (such as was mentioned in the previous paragraph) invariably involve additional issues, such as how one deals with irregularly-spaced data, and (as importantly) the question, are the frequencies that are found real? The Lomb algorithm given in [32] actually addresses both of these matters: it does not require regularly-spaced data, and each spectral peak is accompanied by a probability value. The latter provides the answer to the question, 'if the signal stems from noise, what is the probability that a spectral peak with power W would be observed?" The smaller the probability value, the less likely it is that the peak was randomly-generated[3]. For the Lomb algorithm, it can be shown that the relevant probability distribution is the exponential distribution (see Appendix F); refer to [45] for more information.

---

[2]Using the politically-correct term.
[3]Such statistical significance values occur in other contexts, and may appear in the form of a *p-value*, with essentially the same meaning.

Chapter 17

COMPLEX NUMBERS

ONE might be entitled to ask, "aren't numbers complex enough, without making them any worse!" But the term *complex number* refers to a construct such as

$$z = x + iy,$$

where $i = \sqrt{-1}$ (so that $i^2 = -1$) and x, y are numbers as usual[1]. It took a long time for the quantity i to be accepted as existing at all, thus generating the now-standard (but somewhat unfortunate) nomenclature of x as the real part of z and iy as the imaginary part.

There are similarities here with vectors in two dimensions, x being along one axis (typically horizontal) and y along the other (typically vertical). It is often convenient to visualise complex numbers as points on a plane, as in Figure 2.1. Thus each point can either be addressed as the Cartesian pair (x, y), or in polar form by a distance r and angle θ, such that $z = re^{i\theta}$. That is, in equivalent terms, $x = r\cos\theta$, $y = r\sin\theta$.

Without complex numbers, equations such as $x^2 + 3 = 0$ cannot be solved. Furthermore, in the 1920s, it was found that the dynamics of matter and energy at the smallest scales cannot be adequately described or predicted in the absence of i. An example is Schrödinger's equation:

$$i\hbar \frac{\partial \psi}{\partial t} = H\psi,$$

where $\hbar$ is Planck's constant divided by 2π, H is the Hamiltonian operator and ψ is the wave function [46]. This equation is examined a bit further in Chapter 30.

Recent research work has found that quantum mechanics cannot exist without i [47], so evidently there is something fundamental about it (and, given this, it can perhaps be argued that i is even more fundamental than the familiar real numbers!)

Complex numbers also have application in areas other than quadratic equation solving and quantum mechanics, and two of the more surprising ones are described below.

---

[1] In electromagnetic theory, j is often used to stand for $\sqrt{-1}$, presumably because i or its upper-case equivalent already denotes current.

17.1 Point-In-Polygon

When rendering closed shapes on a computer screen, there is often a need to segregate different regions by colour. By eye, on a sheet of paper, this is straightforward — find the boundary and fill in the enclosed area using a crayon of the right colour. For a computer, this is not so easy, since the computer needs to work out which screen points (pixels) are inside the boundary and which are outside. Various fast algorithms have been devised for this task, but it turns out that one of the more direct stems from Cauchy's theorem[2] [48]. This states that:

$$\frac{1}{2\pi i} \oint_C \frac{f(z)\,dz}{z-a} = \begin{cases} f(a) & \text{if } a \text{ is inside the contour } C, \\ 0 & \text{if } a \text{ is outside } C. \end{cases} \quad (17.1)$$

The closed contour C is traversed in an anti-clockwise manner and z, a are complex numbers. The function f needs to be what is called *analytic* (essentially, differentiable in the complex domain, of which more below) and C should be closed and without crossings.

Equation (17.1) originates from Green's theorem in the plane, and so is connected to equation (13.1) (even though the relationship may not be obvious).

To make use of the above result in determining whether the point a falls inside a closed boundary, set $f(z) = 1$ (as the simplest possible option), and integrate $1/(z-a)$ around the boundary. In the particular case where the boundary is formed from n linear sections, equation (17.1) may be simplified to the following test:

$$\text{if } \left| \text{Im} \sum_{i=1}^{n-1} \log_e \left(\frac{z_{i+1} - a}{z_i - a} \right) \right| > 0, \text{ the point is inside; otherwise it is outside.}$$

The operator 'Im' stands for the imaginary part of the sum and the vertical lines denote magnitude. The quantity inside the summation is readily evaluated using the C complex math library.

Now try this on the boundary of Figure 13.2, generating random points in the outer area and colour-coding each point depending on whether the above equation says that the point is inside the boundary or outside. The results, using 2000 points total, are shown in Figure 17.1.

---

[2] Augustin Cauchy, 1789 to 1857.

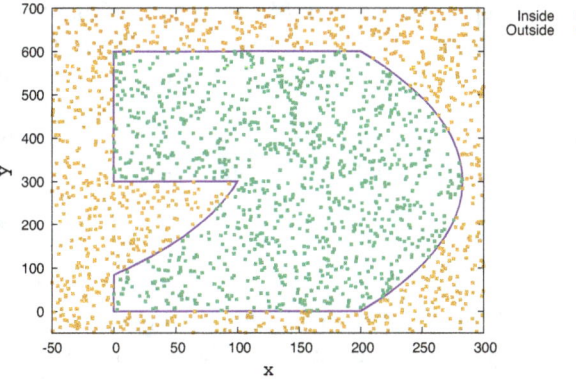

Figure 17.1: Points inside or outside a boundary

As a bonus, generating the random points allows the estimation of the area using the Monte Carlo method described in Section 13.2; with 2000 points, the area comes out to be 140700 square metres, which is not too different from the number stated in Section 13.1.

It may be asked, in equation (17.1), what happens if the the point a just happens to fall bang on the contour C? In this case, the value of the integral will depend on the local nature of the contour — specifically, whether it has a continuous first derivative or not at a. It is not proposed to go any further here into deriving the answer, but the techniques for contour integration in the complex plane can be found in [6] or [48], for example.

17.2 Applications to Fluid Flow and Electrostatics

We kick this section off with a bit more explanation of what constitutes 'analytic', meaning differentiability in the complex domain. Suppose we are dealing with some complex function

$$\Omega(x, y) = u(x, y) + iv(x, y),$$

here subdividing it into real and imaginary parts. Then a sufficient condition for analyticity is that the Cauchy-Riemann equations hold, namely:

$$\frac{\partial u}{\partial x} = \frac{\partial v}{\partial y}, \quad \frac{\partial u}{\partial y} = -\frac{\partial v}{\partial x}.$$

If we now go to the second derivative, we find (try it for yourself) that

$$\frac{\partial^2 \Omega}{\partial x^2} + \frac{\partial^2 \Omega}{\partial y^2} = 0,$$

which is called *Laplace's Equation* for a function in two spatial dimensions[3].

---
[3] Pierre Simon, Marquis de Laplace, 1749 to 1827.

So any analytic function satisfies Laplace's equation. But this same equation also occurs in electrostatics, and in hydrodynamics and aerodynamics (at least under somewhat idealised conditions [48]), and it is the latter connection that will be explored a bit further here.

We start with one of the simplest possible fluid flows, namely linear uninterrupted motion from left to right, for which the complex potential is given by

$$\Omega = V_0 w, \text{ where } w = p + iq.$$

The imaginary part of this, $V_0 q$, provides what are known as *streamlines*, which in experimental situations are made visible by the injection of coloured smoke or dye into the flow. Figure 17.2 provides a visualisation, with the arrows indicating the direction of flow.

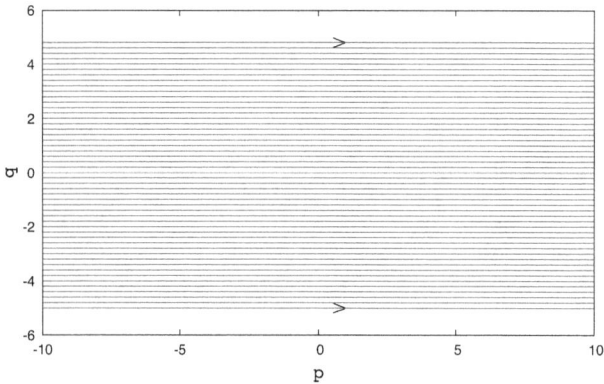

Figure 17.2: Free streamlines

Not very interesting so far, but the fun stuff starts when we provide a mapping from one complex domain to another. Suppose that we shift from w-space to z-space, with

$$w = z + \frac{a^2}{z}, \tag{17.2}$$

with a some real parameter and $z = x + iy$. Equating real and imaginary parts, we have:

$$p = x\left(1 + \frac{a^2}{x^2 + y^2}\right),$$
$$q = y\left(1 - \frac{a^2}{x^2 + y^2}\right),$$

which can be shown to map the whole half-plane $q \geq 0$ to the area greater than $y = 0$ and outside the semicircle $r = \sqrt{x^2 + y^2} = a$. This mapping is illustrated in Figure 17.3.

17.2. APPLICATIONS TO FLUID FLOW AND ELECTROSTATICS

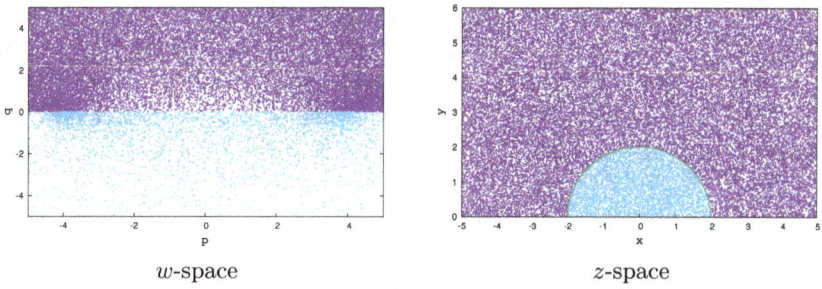

Figure 17.3: Mapping from w-space to z-space

The way these pictures were generated was deliberately made simple: generate a large number of random points in z-space and use the above equations to get p and q; this happens to be an easier route than the converse operation. As an aside, it can be seen that a uniform random distribution on the right corresponds to a non-uniform distribution on the left, with quite obvious concentrations of points in w. The main takeaway point is, though, that the purple domain on the left corresponds to the purple domain on the right.

But so what? The justification for all of this is what happens when the $w \to z$ mapping is used to transform the flow field in Figure 17.2 into the flow over a long cylinder with circular cross-section, and the resulting streamlines for this are given in Figure 17.4. Flow is from left to right and the cylinder has radius 2 units.

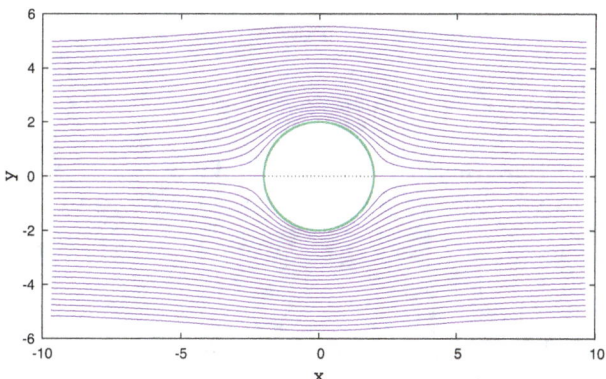

Figure 17.4: Streamlines over a cylinder cross-section

These streamlines are readily derived by inverting equation (17.2) to obtain

$$z = \frac{1}{2}\left(w \pm \sqrt{w^2 - 4a^2}\right).$$

Of the two roots to this equation, the relevant one is that with $|z| \geq a$, since the flow is exterior to the cylinder.

The same mapping-type approach can be used to gain visualisations of the flow over other shapes, or flows around corners, and so forth. It is also possible to derive the flow over an aerofoil-like shape, similar to the cross-section of an aeroplane wing, which is of obvious practical interest. This can be obtained using the following mappings:

$$w = (z - z_0) + \frac{a^2}{(z - z_0)}, \quad \text{which gives flow over a shifted cylinder,}$$

$$\zeta = z + \frac{b^2}{z}, \quad \text{which maps the circle to an aerofoil shape.}$$

The sequence is $w \to z \to \zeta$; in ζ-space, the streamlines now look like Figure 17.5, and the axes correspond to the real and imaginary parts of $\zeta = \mu + i\nu$. Flow is again from left to right, $z_0 = -0.2 + 0.2i$ and $b = 1.8$.

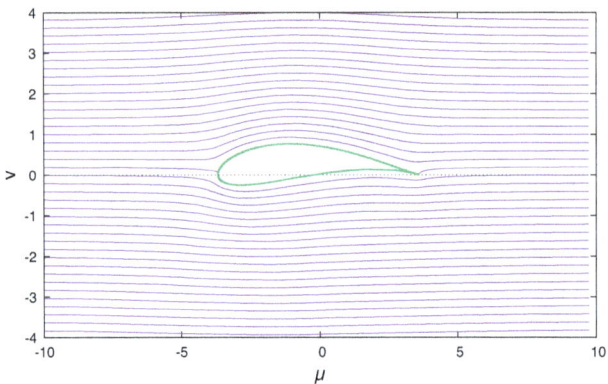

Figure 17.5: Streamlines over an aeroplane wing

As a final remark concerning the practicality of all this, there is a theorem due to Blasius[4] which gives the net fluid force on the obstacle as

$$F = \frac{1}{2} i\rho \oint_C \left(\frac{d\Omega}{dz}\right)^2 dz,$$

where ρ is the fluid density, Ω is the complex potential and the integration is over the surface contour. This can be used to derive aerodynamic lift, for instance.

17.3 Euler's Formula

Take a look back now at the series expansion for the exponential function in Section 7.2, namely

$$e^x = 1 + x + \frac{x^2}{2!} + \frac{x^3}{3!} + \frac{x^4}{4!} + \frac{x^5}{5!} + \cdots,$$

---
[4]Paul Blasius, 1883 to 1970.

17.4. SUMMATION OF SERIES

and set $x = i\theta$, with θ a real quantity. So,

$$e^{i\theta} = 1 + i\theta + \frac{(i\theta)^2}{2!} + \frac{(i\theta)^3}{3!} + \frac{(i\theta)^4}{4!} + \ldots,$$

$$= 1 + i\theta - \frac{\theta^2}{2!} - i\frac{\theta^3}{3!} + \frac{\theta^4}{4!} + i\frac{\theta^5}{5!} \ldots, \text{ since } i^2 = -1, i^3 = i, \text{ etc,}$$

$$= \left(1 - \frac{\theta^2}{2!} + \frac{\theta^4}{4!} - \ldots\right) + i\left(\theta - \frac{\theta^3}{3!} + \frac{\theta^5}{5!} - \ldots\right), \text{ on collecting terms.}$$

Now compare this with the series expansions for the sine and cosine functions in Section 7.3. Evidently,

$$e^{i\theta} = \cos\theta + i\sin\theta, \quad (17.3)$$

which is called *Euler's formula*.

This links the exponential function in the complex domain with the trigonometric functions in the real domain — interesting enough in itself. But set $\theta = \pi$; then since $\sin\pi = 0$ and $\cos\pi = -1$, we have (after some rearrangement of terms),

$$e^{i\pi} + 1 = 0.$$

This short equation links five of the most important constants in mathematics: zero, unity, π, e and i. It is quite extraordinary — there is no earthly reason why raising e (an irrational number) to an imaginary power involving π (another irrational number) should be identical to -1. But it is so!

Return to equation (17.3) and work out half the sum of $e^{i\theta}$ and $e^{-i\theta}$; clearly,

$$\cos\theta = \frac{1}{2}\left(e^{i\theta} + e^{-i\theta}\right). \quad (17.4)$$

And in a similar manner,

$$\sin\theta = \frac{1}{2i}\left(e^{i\theta} - e^{-i\theta}\right). \quad (17.5)$$

These will come in handy shortly.

17.4 Summation of Series

Complex numbers can be used to evaluate expressions that would otherwise be distinctly opaque. For example, suppose we have the series

$$S_n = \sum_{k=0}^{n-1} \cos kx.$$

Although this particular instance can be found in tables of integrals and series (*e.g.* [36]), it is still instructive to be aware of at least one evaluative method as a back-up, just in case the search for a standard result comes up blank.

In view of equation (17.3), write the cosine function as the real part of a complex exponential, so that

$$\cos kx = Re\left[e^{ikx}\right],$$

using the operator Re to stand for the real part (and we would use Im for the imaginary bit, although this isn't needed here). Therefore,

$$S_n = Re\left[\sum_{k=0}^{n-1} e^{ikx}\right],$$

since Re is a linear operator.

Write this out in full:

$$S_n = Re\left[1 + e^{ix} + e^{2ix} + e^{3ix} + \ldots + e^{i(n-1)x}\right].$$

If this looks suspiciously like the geometric series that was touched on in Section 4.1, that's because it is. So we can immediately determine that

$$S_n = Re\left[\frac{1 - e^{inx}}{1 - e^{ix}}\right].$$

This is still in complex form, while ideally we would like to remove i from the expression. This is straightforward enough: reverse the ordering of both top and bottom, and multiply top and bottom by $e^{-ix/2}$:

$$S_n = Re\left[\frac{e^{-ix/2}\left(e^{inx} - 1\right)}{e^{-ix/2}\left(e^{ix} - 1\right)}\right],$$

$$= Re\left[\frac{\left(e^{i(n-1/2)x} - e^{-ix/2}\right)}{\left(e^{ix/2} - e^{-ix/2}\right)}\right], \text{ after multiplying out,}$$

$$= Re\left[\frac{\left(e^{i(n-1/2)x} - e^{-ix/2}\right)}{2i \sin x/2}\right], \text{ from equation (17.5),}$$

$$= Re\left[\frac{-i\left(e^{i(n-1/2)x} - e^{-ix/2}\right)}{2 \sin x/2}\right], \text{ multiplying top and bottom by } i.$$

If the numerator is now expanded back out into $\cos + \sin$ form and the real part extracted, we eventually get

$$S_n = \frac{1}{2}\left[1 + \frac{\sin(n - 1/2)x}{\sin x/2}\right].$$

The complex numbers have here acted somewhat like a catalyst: entering an equation, carrying out the necessary work and then quietly bowing out at the end.

The same method works for summations of the form $\sum_k \cos^2 kx, \sum_k \sin^2 kx$, making use of the relationships $\cos 2\theta = 2\cos^2\theta - 1 = 1 - 2\sin^2\theta$.

Chapter 18

QUATERNIONS

I must confess to having had little to do with quaternions during my working life, except for one instance when it was necessary to convert a satellite's observational pointing vector from body coordinates into earth-relative ones (and then the necessary calculations were included in the accompanying documentation). And yet quaternions have proved very popular in the robotics and computer graphics industries (among other applications in physics), so a brief discussion is warranted here.

Quaternions share some of their behavioural characteristics with complex numbers and some characteristics with ordinary three-dimensional vectors, thus requiring their own specific operational algebra. The investigation of these entities commenced in 1843 by Hamilton[1], who was looking to extend the concept of rotation in the two-dimensional complex plane into three space dimensions. To clarify the latter statement, suppose we have a complex number $z = x + iy$, with $i^2 = -1$, and regard it as a vector in the plane (see Section 17). Then write it in the polar form $z = re^{i\theta}$, and multiply it by the unit complex number $e^{i\phi}$, to give $re^{i(\theta+\phi)}$. The vector will then have been rotated by the angle ϕ (and with unchanged length).

Hamilton eventually found that the following quantity satisfied his search requirements:

$$q = q_0 + q_1\mathbf{i} + q_2\mathbf{j} + q_3\mathbf{k}, \tag{18.1}$$

in which the q_i are scalars (real numbers in the usual sense of the word), while the $\mathbf{i}, \mathbf{j}, \mathbf{k}$ entities are the familiar unit orthonormal vectors, namely $\mathbf{i} = (1, 0, 0)\,; \mathbf{j} = (0, 1, 0)\,; \mathbf{k} = (0, 0, 1)$, but having the additional multiplicative properties:

$$\mathbf{i}^2 = \mathbf{j}^2 = \mathbf{k}^2 = \mathbf{ijk} = -1, \tag{18.2}$$
$$\mathbf{ij} = -\mathbf{ji} = \mathbf{k}, \;\; \mathbf{jk} = -\mathbf{kj} = \mathbf{i}, \;\; \mathbf{ki} = -\mathbf{ik} = \mathbf{j}. \tag{18.3}$$

Equation set (18.2) is reminiscent of complex numbers, while equation set (18.3) harks back toward the usual vector products, as in $\mathbf{i} \wedge \mathbf{j} = \mathbf{k}$ (see Chapter 14).

---

[1] William Rowan Hamilton, 1805 to 1865.

It may be observed from the above that quaternion algebra is not commutative in multiplication, a property shared with matrices and operators. Addition and subtraction of quaternions operates component-wise, in a similar manner to ordinary numbers and vectors.

The analogue of the complex conjugate also appears with with quaternions, so that

$$q^* = q_0 - q_1\mathbf{i} - q_2\mathbf{j} - q_3\mathbf{k},$$

and it is often convenient to write

$$q = q_0 + \mathbf{q}, \text{ with } \mathbf{q} = q_1\mathbf{i} + q_2\mathbf{j} + q_3\mathbf{k},$$

thus separating a quaternion into its scalar and vector (or 'imaginary') parts.

A key result that will be needed shortly is that

$$qq^* = q^*q = q_0^2 + q_1^2 + q_2^2 + q_3^2.$$

This follows directly from the multiplicative properties in equations (18.2) and (18.3), but it is still instructive to follow through the details:

$$qq^* = (q_0 + \mathbf{q})(q_0 - \mathbf{q}) = q_0^2 - q_0\mathbf{q} + q_0\mathbf{q} - \mathbf{qq} = q_0^2 - \mathbf{qq}.$$

But, multiplying out term by term,

$$\mathbf{qq} = (q_1\mathbf{i} + q_2\mathbf{j} + q_3\mathbf{k})(q_1\mathbf{i} + q_2\mathbf{j} + q_3\mathbf{k}),$$
$$= q_1^2\mathbf{i}^2 + q_1q_2\mathbf{ij} + q_1q_3\mathbf{ik} + q_1q_2\mathbf{ji} + q_2^2\mathbf{j}^2 + q_2q_3\mathbf{jk} + q_1q_3\mathbf{ki}$$
$$+ q_2q_3\mathbf{kj} + q_3^2\mathbf{k}^2,$$
$$= -q_1^2 - q_2^2 - q_3^2,$$

since $\mathbf{i}^2 = -1$, etc, and $\mathbf{ij} = -\mathbf{ji}$, etc, from the definitions in equations (18.2) and (18.3).

We will also need the product of two quaternions[2] p and q, and going through the details in the same manner highlights a useful abbreviated form for the solution:

$$qp = (q_0 + q_1\mathbf{i} + q_2\mathbf{j} + q_3\mathbf{k})(p_0 + p_1\mathbf{i} + p_2\mathbf{j} + p_3\mathbf{k}),$$
$$= q_0p_0 + q_0p_1\mathbf{i} + q_0p_2\mathbf{j} + q_0p_3\mathbf{k} + q_1p_0\mathbf{i} + q_1p_1\mathbf{i}^2 + q_1p_2\mathbf{ij} + q_1p_3\mathbf{ik}$$
$$+ q_2p_0\mathbf{j} + q_2p_1\mathbf{ji} + q_2p_2\mathbf{j}^2 + q_2p_3\mathbf{jk} + q_3p_0\mathbf{k} + q_3p_1\mathbf{ki}$$
$$+ q_3p_2\mathbf{kj} + q_3p_3\mathbf{k}^2,$$
$$= q_0p_0 - (q_1p_1 + q_2p_2 + q_3p_3) + q_0\mathbf{p} + p_0\mathbf{q}$$
$$+ \mathbf{i}(q_2p_3 - q_3p_2) + \mathbf{j}(q_3p_1 - q_1p_3) + \mathbf{k}(q_1p_2 - q_2p_1).$$

---

[2] Multiplication of quaternions is sometimes written in the form $p \otimes q$, but provided we keep in mind that p and q are quaternions there should be no confusion with using the simpler form pq.

Now refer back to the vector scalar and cross product definitions in Chapter 14; it is then found that the product qp simplifies to the following:

$$qp = q_0 p_0 - \mathbf{q} \cdot \mathbf{p} + q_0 \mathbf{p} + p_0 \mathbf{q} + \mathbf{q} \wedge \mathbf{p}, \tag{18.4}$$

in which $\mathbf{p}$ and $\mathbf{q}$ stand for the vector parts of their respective quaternions.

We now have the necessary apparatus to show how quaternions can be used to rotate a vector through any specific angle.

Suppose the vector $\mathbf{q}$ is itself defined in terms of a unit vector, so that

$$\mathbf{q} = \lambda \hat{\mathbf{a}}, \text{ with } |\hat{\mathbf{a}}| = 1.$$

Then if quaternion q is defined as follows:

$$q = q_0 + \lambda \hat{\mathbf{a}}, \text{ with } q_0 = \cos(\theta/2) \text{ and } \lambda = \sin(\theta/2), \tag{18.5}$$

we find that the construct

$$qvq^*$$

will rotate any three-component vector $\mathbf{v}$ around axis $\hat{\mathbf{a}}$ through the angle θ. This will be demonstrated below.

For convenience in manipulating the various operations that will be needed, suppose we define the following rotational operator $\mathcal{R}$:

$$\mathcal{R}\mathbf{v} \equiv qvq^*,$$

and apply this to a vector sum:

$$\begin{aligned}
\mathcal{R}(\mathbf{v}+\mathbf{w}) &= q(\mathbf{v}+\mathbf{w})q^*, \\
&= q(\mathbf{v}q^* + \mathbf{w}q^*), \text{ multiplying out term by term,} \\
&= q\mathbf{v}q^* + q\mathbf{w}q^*, \\
&= \mathcal{R}\mathbf{v} + \mathcal{R}\mathbf{w},
\end{aligned}$$

which means that $\mathcal{R}$ is a linear operator.

Correctly, we should be a bit more specific about the constituents of the operator $\mathcal{R}$ and write it in the form $\mathcal{R}(\hat{\mathbf{a}}, \theta)$, so that it is then clear what axis and angle are involved, but the abbreviated notation will do for now.

Next have a look at what $\mathcal{R}$ does to a vector $\mathbf{v}$ that happens to be aligned with the $\hat{\mathbf{a}}$ in equation (18.5), meaning that $\mathbf{v} = \alpha \hat{\mathbf{a}}$ for some scalar parameter α. Plug this $\mathbf{v}$ into equation (18.4) in place of p, using $p_0 = 0$ (thus just treating $\mathbf{v}$ as a quaternion with a zero scalar part):

$$q\mathbf{v} = -(\lambda \hat{\mathbf{a}}) \cdot (\alpha \hat{\mathbf{a}}) + q_0(\alpha \hat{\mathbf{a}}) + (\lambda \hat{\mathbf{a}}) \wedge (\alpha \hat{\mathbf{a}}) = -\lambda \alpha + q_0 \alpha \hat{\mathbf{a}}, . \tag{18.6}$$

since $\hat{\mathbf{a}} \cdot \hat{\mathbf{a}} = 1$ and $\hat{\mathbf{a}} \wedge \hat{\mathbf{a}} = 0$ (see Chapter 14). Go back to equation (18.4), and reverse the product ordering, so that

$$pq = p_0 q_0 - \mathbf{p} \cdot \mathbf{q} + p_0 \mathbf{q} + q_0 \mathbf{p} + \mathbf{p} \wedge \mathbf{q},$$

and then replace q by its complex conjugate (which just means reversing the sign of $\mathbf{q}$ in the expression), resulting in

$$pq^* = p_0 q_0 + \mathbf{p} \cdot \mathbf{q} - p_0 \mathbf{q} + q_0 \mathbf{p} - \mathbf{p} \wedge \mathbf{q}. \tag{18.7}$$

Given the right-hand-side of equation (18.6) as p, we must have $p_0 = -\lambda\alpha$, $\mathbf{p} = q_0 \alpha \hat{\mathbf{a}}$, while q_0 and $\mathbf{q} = \lambda \hat{\mathbf{a}}$ are as before. Therefore, substituting into pq^*:

$$\begin{aligned}
qvq^* &= p_0 q_0 + \mathbf{p} \cdot \mathbf{q} - p_0 \mathbf{q} + q_0 \mathbf{p} - \mathbf{p} \wedge \mathbf{q}, \\
&= (-\lambda\alpha) q_0 + (q_0 \alpha \hat{\mathbf{a}}) \cdot (\lambda \hat{\mathbf{a}}) - (-\lambda\alpha)(\lambda \hat{\mathbf{a}}) + q_0 (q_0 \alpha \hat{\mathbf{a}}) - (q_0 \alpha \hat{\mathbf{a}}) \wedge (\lambda \hat{\mathbf{a}}), \\
&= \alpha \left(q_0^2 + \lambda^2 \right) \hat{\mathbf{a}}, \\
&= \alpha \hat{\mathbf{a}}, \text{ since } q_0^2 + \lambda^2 = 1 \text{ from their definitions in equation (18.5)}, \\
&= \mathbf{v}.
\end{aligned}$$

So if $\mathbf{v}$ is aligned with the $\hat{\mathbf{a}}$ in the vector part of q, the operation $\mathcal{R}\mathbf{v}$ leaves $\mathbf{v}$ unchanged in magnitude and direction. It may be inferred that any rotation must be about the $\hat{\mathbf{a}}$ axis.

For a more general vector $\mathbf{v}$, decompose it into components along and normal to $\hat{\mathbf{a}}$, such that $\mathbf{v} = \alpha \hat{\mathbf{a}} + \beta \hat{\mathbf{s}} + \gamma \hat{\mathbf{t}}$, and write $\mathbf{c} = \beta \hat{\mathbf{s}} + \gamma \hat{\mathbf{t}}$, so that $\mathbf{v} = \alpha \hat{\mathbf{a}} + \mathbf{c}$, noting that $\mathbf{c} \perp \hat{\mathbf{a}}$. This decomposition is valid since we are free to choose the orthonormal basis vector set. Then examine $\mathcal{R}(\alpha \hat{\mathbf{a}} + \mathbf{c})$ and concentrate on the $\mathcal{R}\mathbf{c}$ component, as the $\alpha \hat{\mathbf{a}}$ part is unchanged by the rotation. That is, look at

$$\begin{aligned}
q\mathbf{c} &= -(\lambda \hat{\mathbf{a}}) \cdot \mathbf{c} + q_0 \mathbf{c} + (\lambda \hat{\mathbf{a}}) \wedge \mathbf{c}, \\
&= q_0 \mathbf{c} + \lambda \hat{\mathbf{a}} \wedge \mathbf{c}, \text{ since } \hat{\mathbf{a}} \text{ is normal to } \mathbf{c}.
\end{aligned}$$

Bear in mind that $\mathbf{a} \wedge \mathbf{c}$ is perpendicular to both $\hat{\mathbf{a}}$ and $\mathbf{c}$.

We next need equation (18.7), although now with $p_0 = 0$ and $\mathbf{p} = q_0 \mathbf{c} + \lambda \hat{\mathbf{a}} \wedge \mathbf{c}$; again, q_0 and $\mathbf{q} = \lambda \hat{\mathbf{a}}$ are as before. Then,

$$\begin{aligned}
q\mathbf{c}q^* &= (q_0 \mathbf{c} + \lambda \hat{\mathbf{a}} \wedge \mathbf{c}) \cdot (\lambda \hat{\mathbf{a}}) + q_0 (q_0 \mathbf{c} + \lambda \hat{\mathbf{a}} \wedge \mathbf{c}) - (q_0 \mathbf{c} + \lambda \hat{\mathbf{a}} \wedge \mathbf{c}) \wedge (\lambda \hat{\mathbf{a}}), \\
&= q_0^2 \mathbf{c} + \lambda q_0 \hat{\mathbf{a}} \wedge \mathbf{c} - \lambda q_0 \mathbf{c} \wedge \hat{\mathbf{a}} - \lambda^2 \hat{\mathbf{a}} \wedge \mathbf{c} \wedge \hat{\mathbf{a}}, \\
&= q_0^2 \mathbf{c} + 2\lambda q_0 (\hat{\mathbf{a}} \wedge \mathbf{c}) - \lambda^2 (\hat{\mathbf{a}} \wedge \mathbf{c}) \wedge \hat{\mathbf{a}}. \tag{18.8}
\end{aligned}$$

Here, the first term on the first line vanishes since $\hat{\mathbf{a}}$ is normal to both $\mathbf{c}$ and to $\mathbf{c} \wedge \hat{\mathbf{a}}$. Also, $\mathbf{c} \wedge \hat{\mathbf{a}} = -\hat{\mathbf{a}} \wedge \mathbf{c}$ has been used to simplify the second line[3].

At this stage, it helps to revert back to writing $\mathbf{c} = \beta \hat{\mathbf{s}} + \gamma \hat{\mathbf{t}}$, where $\hat{\mathbf{a}}, \hat{\mathbf{s}}, \hat{\mathbf{t}}$ form a right-handed orthonormal vector set. That is [9],

$$\hat{\mathbf{a}} \wedge \hat{\mathbf{s}} = \hat{\mathbf{t}}; \quad \hat{\mathbf{t}} \wedge \hat{\mathbf{a}} = \hat{\mathbf{s}}; \quad \hat{\mathbf{s}} \wedge \hat{\mathbf{t}} = \hat{\mathbf{a}}.$$

Therefore,

$$\begin{aligned}
\hat{\mathbf{a}} \wedge \mathbf{c} &= \hat{\mathbf{a}} \wedge (\beta \hat{\mathbf{s}} + \gamma \hat{\mathbf{t}}) = \beta \hat{\mathbf{t}} - \gamma \hat{\mathbf{s}}, \\
(\hat{\mathbf{a}} \wedge \mathbf{c}) \wedge \hat{\mathbf{a}} &= (\beta \hat{\mathbf{t}} - \gamma \hat{\mathbf{s}}) \wedge \hat{\mathbf{a}} = \beta \hat{\mathbf{s}} + \gamma \hat{\mathbf{t}}.
\end{aligned}$$

---

[3] Recall that the vector product is anti-commutative.

Plug these into equation (18.8):

$$qcq^* = q_0^2 \left(\beta\hat{\mathbf{s}} + \gamma\hat{\mathbf{t}}\right) + 2\lambda q_0 \left(\beta\hat{\mathbf{t}} - \gamma\hat{\mathbf{s}}\right) - \lambda^2 \left(\beta\hat{\mathbf{s}} + \gamma\hat{\mathbf{t}}\right),$$
$$= \hat{\mathbf{s}}\left[\beta\left(q_0^2 - \lambda^2\right) - 2\lambda q_0\gamma\right] + \hat{\mathbf{t}}\left[\gamma\left(q_0^2 + \lambda^2\right) + 2\lambda q_0\beta\right].$$

Now make use of the original definitions: $q_0 = \cos(\theta/2)$, $\lambda = \sin(\theta/2)$. Thus, $q_0^2 - \lambda^2 = \cos\theta$ and $2\lambda q_0 = \sin\theta$ (see equations (3.8) and (3.7)), resulting in:

$$qcq^* = \hat{\mathbf{s}}\left[\beta\cos\theta - \gamma\sin\theta\right] + \hat{\mathbf{t}}\left[\gamma\cos\theta + \beta\sin\theta\right],$$
$$= \beta\left(\cos\theta\,\hat{\mathbf{s}} + \sin\theta\,\hat{\mathbf{t}}\right) + \gamma\left(\cos\theta\,\hat{\mathbf{t}} - \sin\theta\,\hat{\mathbf{s}}\right).$$

So the result of the operation $\mathcal{R}\mathbf{c}$ has been to change $\mathbf{c} = \beta\hat{\mathbf{s}} + \gamma\hat{\mathbf{t}}$ into $\mathbf{c}' = \beta\hat{\mathbf{s}}' + \gamma\hat{\mathbf{t}}'$, where

$$\hat{\mathbf{s}}' = \cos\theta\,\hat{\mathbf{s}} + \sin\theta\,\hat{\mathbf{t}} \text{ and } \hat{\mathbf{t}}' = \cos\theta\,\hat{\mathbf{t}} - \sin\theta\,\hat{\mathbf{s}}.$$

This is just a rotation of axes about $\hat{\mathbf{a}}$, changing $\hat{\mathbf{s}}$ and $\hat{\mathbf{t}}$ into $\hat{\mathbf{s}}'$ and $\hat{\mathbf{t}}'$.

If we now revert back to a generic vector $\mathbf{v} = \alpha\hat{\mathbf{a}} + \beta\hat{\mathbf{s}} + \gamma\hat{\mathbf{t}}$, the quaternion-based rotational operation $\mathbf{v}' = \mathcal{R}(\hat{\mathbf{a}}, \theta)\mathbf{v}$ can alternatively be written using the (perhaps more familiar) rotation matrix form:

$$\mathbf{v}' = \Omega\mathbf{v}, \text{ in which } \Omega = \begin{pmatrix} 1 & 0 & 0 \\ 0 & \cos\theta & \sin\theta \\ 0 & -\sin\theta & \cos\theta \end{pmatrix},$$

and the vector component ordering here is defined to be $\hat{\mathbf{a}}, \hat{\mathbf{s}}, \hat{\mathbf{t}}$ for both columns and rows.

Quaternions have proved to be a very efficient means for carrying out rotations in three-dimensions, with less arithmetic operations needed than with the use of rotation matrices. This latter claim may seem a bit controversial, given the compactness of the above $\Omega\mathbf{v}$ operations, but in general $\mathbf{v}$ will be specified in terms of some other coordinate axes ($\mathbf{i}, \mathbf{j}, \mathbf{k}$, say), rather than relative to $\hat{\mathbf{a}}, \hat{\mathbf{s}}, \hat{\mathbf{t}}$. Thus $\mathbf{v}$ needs to be mapped to the $\hat{\mathbf{a}}, \hat{\mathbf{s}}, \hat{\mathbf{t}}$ set before the above $\Omega\mathbf{v}$ can be computed, and the result then converted back to $\mathbf{i}, \mathbf{j}, \mathbf{k}$ afterwards. Whereas qvq^* operates entirely in whatever vector basis is provided. My own numerical experiments indicate that the quaternion route takes about one-quarter of the processing time needed for rotation matrices (hardly a definitive result, as my C code is unlikely to be the most efficient possible, but indicative nonetheless).

For further reading and applications, see (for example) [49], [50], [51].

Chapter 19

POPULATION DYNAMICS

THE recent coronavirus pandemic (2019 onwards) has highlighted the value of mathematical modelling of the spread of infections. The basic approach is straightforward: subdivide the population into groups such as 'susceptible', 'infected', 'recovered' and 'deceased', and then define differential equations describing the interactions between these groups. Comprehensive models of this nature can be found in [52] and [53], but the general idea is more easily appreciated by looking initially at the *Lotka-Volterra* equations[1] describing predator-prey dynamics, here concentrating on foxes and rabbits[2]. The slightly more complex equations for epidemics are then described in Section 19.2.

19.1 Predator-Prey Dynamics

Suppose x describes the number of rabbits, y the number of foxes and t denotes time. Then the time variations of both populations can be described by the following two linked differential equations:

$$\frac{dx}{dt} = \alpha x - \beta xy, \tag{19.1}$$

$$\frac{dy}{dt} = \gamma xy - \mu y, \tag{19.2}$$

where the quantities α, β, γ and μ are modelling parameters.

To understand the basis of these equations, suppose $\beta = 0$ in equation (19.1). Then x — the number of rabbits — will increase exponentially (a well-known rabbit tendency in the absence of any population checks). The βxy and γxy terms describe 'interactions' and are proportional to the expected number of fox-rabbit encounters — the more rabbits and foxes there are around, the more likely they are to meet (even if the rabbits would prefer otherwise). In equation (19.1),

---

[1] Alfred Lotka, 1880 to 1949 and Vito Volterra, 1860 to 1940.
[2] Or lynxes and hares, if preferred.

such interactions reduce the rabbit numbers, while in equation (19.2), the foxes benefit. Similarly, if $\gamma = 0$, the number of foxes decays to zero, since they cannot find rabbits and will run out of food.

As an aside, returning briefly to pandemic modelling, the parameters α, β, γ and μ — or their equivalents in the more complex models — offer opportunities for exploring the impact of interventions such as isolation, vaccinations and so forth. We will come back to that in the next section.

Figure 19.1 shows a typical example of the growth and decay of rabbit and fox numbers with time, commencing with 10 rabbits and 2 foxes. The parameters used were: $\alpha = 1.1$, $\beta = 0.08$, $\gamma = 0.02$ and $\mu = 0.5$. The time units are arbitrary.

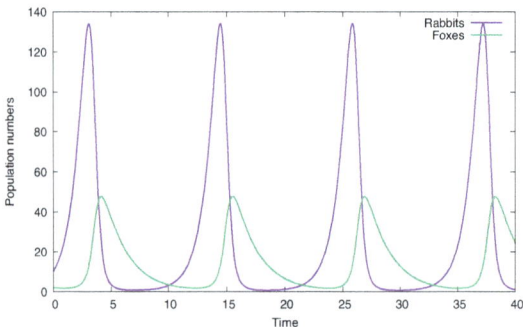

Figure 19.1: Numbers of rabbits and foxes versus time

Evidently the behaviour of both populations is periodic, with no stable state reached. The rabbit population initially grows, followed a short while later by an increase in the fox population. When the foxes have increased sufficiently to make an impact, the rabbit numbers suffer a near-collapse, followed some while later by the fox numbers. And then the cycle starts over again; such behaviours are not unknown in nature, although are unlikely to be as neat as in Figure 19.1.

A different way of looking at the population dynamics is to suppress the time coordinate and plot y versus x, as in Figure 19.2.

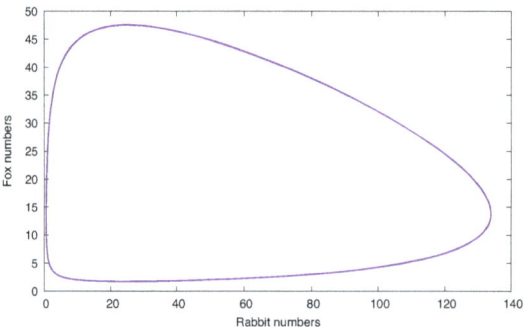

Figure 19.2: Fox numbers versus rabbit numbers

19.2. EPIDEMICS

This collapses endless cycles of growth and decay into a single closed curve and its equation can be derived from equation (19.2) divided by (19.1), to eliminate the time coordinate (which in Figure (19.2) runs around the curve):

$$\frac{dy}{dx} = \frac{dy/dt}{dx/dt} = \frac{y(\gamma x - \mu)}{x(\alpha - \beta y)}.$$

Solving this gives:

$$\alpha \log_e y + \mu \log_e x + C = \gamma x + \beta y,$$

where C is a constant to be set from the initial conditions.

Equations (19.1) and (19.2) are, of course, idealisations. In reality, interactions with other species would take place and random effects would also need to be taken into account. Nonetheless, insights into the overall behaviour can still be gained from such simple models.

19.2 Epidemics

Here, we restrict ourselves to the SIR — Susceptible-Infected-Recovered — model, in which the relevant population is divided into those three groups. The absence of a 'deceased' category means that the infection is assumed to be non-fatal. The following differential equations then describe the time variation of the numbers in each group [54]:

$$\frac{ds}{dt} = -\beta si, \qquad (19.3)$$

$$\frac{di}{dt} = \beta si - \nu i, \qquad (19.4)$$

$$\frac{dr}{dt} = \nu i. \qquad (19.5)$$

The s (susceptible), i (infected) and r (recovered) quantities represent the relevant proportions of the population, rather than absolute numbers, so $s+i+r=1$. Since these groups form part of a wider whole, the same interaction term βsi appears in the differential equations for both s and i, but with opposite signs.

The above population is assumed to be stable in size (no births or deaths), and well mixed so that interaction rates are representative across all personnel.

The parameter β, here assumed constant, is defined as

$$\beta = \tau c,$$

where τ is the transmissibility coefficient and c is the average rate of contact between susceptible and infected people.

Also, ν (again, assumed constant) is the recovery rate of infected people, and can be related to the duration of infectiousness d via

$$\nu = \frac{1}{d}.$$

A key parameter that has received a lot of recent attention in the media is the *reproduction number R*: the average number of people each infected person will go on to infect in turn. For the above simple model, this is defined as:

$$R = \tau c d = \beta d = \frac{\beta}{\nu}. \tag{19.6}$$

More complex models come with more complicated definitions of R [52], [53].

Useful information can be obtained from equation (19.4), even without attempting to solve for s and i. Write the equation in the form

$$\frac{di}{dt} = i(\beta s - \nu).$$

Since this describes the rate of change of i, it can be inferred that the number of infections will grow with time if $\beta s > \nu$ and will decay otherwise. That is, growth of an epidemic if

$$s > \frac{\nu}{\beta}, \text{ which may also be written as } s > \frac{1}{R}.$$

Thus, right at the start of a potential epidemic, it can be determined whether or not it will 'take off' based on the initial proportion of the susceptible population plus the constants β and ν. So if a sufficient number of people have been vaccinated, for example, an epidemic can be prevented at the outset. This concept of what is often termed 'herd immunity' (a vaguely rude term, to my mind) is discussed further toward the end of this section.

Representative variations over time for each of the three population categories is shown in Figure 19.3, assuming $R = 2$ and $d = 5$ days. Initially, out of a population of 10,000 people, only two were presumed infected.

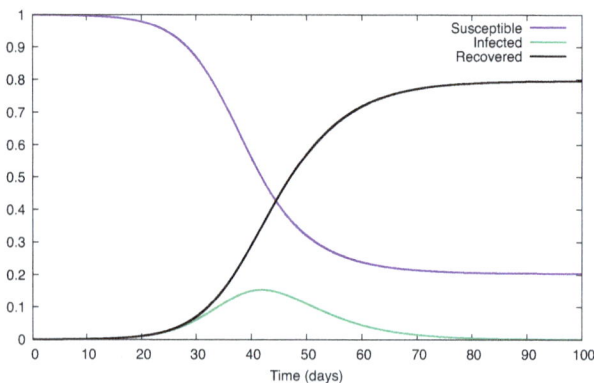

Figure 19.3: Susceptible, infected and recovered proportions of the population

In contrast to the fox-rabbit simulation in Section 19.1, which exhibited periodic behaviour, a stable state is reached here after about 80 days. The infection increases, reaches a peak at about mid-way through, and then decays again — pretty much consistent with experience of infectious diseases.

19.2. EPIDEMICS

Another aspect of Figure 19.3 that is noticeable is that the number of susceptible people does not tend to zero with increasing time, but reaches a stable *non*-zero state. This means that it is not necessary for everyone to be immune for the infection to eventually die out — a useful, if unexpected, prediction that is another manifestation of herd immunity.

To see why $s > 0$ as $t \to \infty$, it helps to have a single equation linking s and i. It may be noticed that equations (19.3) and (19.4) are independent of r. So following the approach in Section 19.1, divide one differential equation by the other to obtain:

$$\frac{ds}{di} = -\frac{\beta s}{\beta s - \nu}.$$

This is readily integrated to give

$$s - \frac{\nu}{\beta} \log_e s = C - i,$$

where C is a constant. Using a suffix $_0$ to indicate 'initial value', it is found that

$$C = i_0 + s_0 - \frac{\nu}{\beta} \log_e s_0,$$

$$= 1 - \frac{\nu}{\beta} \log_e s_0, \text{ since } r_0 = 0 \text{ by definition and } i + s + r = 1.$$

There is an implied constraint here, in that $s_0 \neq 0$, as otherwise C becomes unbounded unless $\nu/\beta = 0$ also.

We thus obtain a single equation linking s and i:

$$s + \frac{1}{R} \log_e \left(\frac{s_0}{s}\right) = 1 - i,$$

since $R = \beta/\nu$ in this model. Rearrange this to put i on the left, so that

$$i = 1 - s - \frac{1}{R} \log_e \left(\frac{s_0}{s}\right).$$

Now ask what would happen if s really tended to zero, meaning no susceptible people as $t \to \infty$. The log term then blows up, since $s_0 > 0$, and $i \to -\infty$, which is nonsense[3]. So it is expected that $s > 0$, even as $t \to \infty$.

It was mentioned above that preventing an infection from taking hold in a population would require that

$$s_0 < \frac{1}{R}.$$

It will be noted that if $R < 1$, this condition is satisfied automatically, so the subsequent discussion is restricted to the situation where $R > 1$, which is usually of more concern.

---

[3] Or non-physical, to put it a bit more formally.

To relate the above inequality to the 'herd immunity' concept, introduce a converse grouping: the initial proportion of the population that is immune and call this q_0. Since $s_0 + q_0 = 1$, we must have

$$q_0 > 1 - \frac{1}{R},$$

for the initial infection to die out[4]. The quantity q_0 is the proportion of the population required for 'herd immunity'. So with the simulation parameters used to generate Figure 19.3, giving $R = 2$, at least 50% of the population would need to be immune.

For comparison, consider measles, with R between 12 or 18 [55], or polio with R in the range 4 to 6 [56]. It may then be concluded that for measles, at least 92% of the population needs to be immune, and for polio, 75% to 83%. These figures, predicted by the above simple model, tally quite well with current medical recommendations (93% - 95% and 80% - 86% respectively [57], [58]).

Thus even this simple model provides insights into the progress of an infection, besides suggesting suitable intervention possibilities. For example, the rapid deployment of vaccination measures during an epidemic means that the infection rates can be tamed even if the current value of R implies a worrying situation.

As mentioned at the start of this section, there is no problem in adding extra population sub-groups into the model to accommodate the actual nature of a disease — a 'deceased' category being the most obvious, but also (for example) 'infected but not currently infectious to others'. And so on.

---

[4]Since inequalities can be rather troublesome to deal with, the actual derivation of this result uses the following series of operations: $s_0 + q_0 = 1$ and $s_0 < 1/R$, which together imply that $1 - q_0 < 1/R \Rightarrow 1 < 1/R + q_0 \Rightarrow q_0 > 1 - 1/R$. Adding or subtracting quantities on both sides in order to rearrange an inequality equation is non-contentious, but care needs to be taken when *multiplying* both sides by some quantity. Try it and see what happens. For further information on inequalities, see (for example) [9].

Chapter 20

INTEGRAL TRANSFORMS

An *Integral Transform* converts from one mathematical space into another, which may seem a bit pointless — except that in many cases, a solution may be readily obtained in one of those spaces, but be more difficult in the other. For instance, the Laplace and Fourier Transforms (Chapter 16) act to map an ordinary differential equation to an algebraic one, or a partial differential equation into an ordinary differential equation (the latter being generally more tractable).

Another form of integral transform that has proved invaluable in medical imaging is the Radon Transform[1], and both this and the Laplace Transform are briefly examined in the following two sections.

20.1 Laplace Transform

The Laplace Transform is defined as follows:

$$F(s) = \int_{t=0}^{\infty} e^{-st} f(t) dt.$$

This maps a function $f(t)$ in the time domain into the function $F(s)$ in the transform space, by means of the intermediary function e^{-st}; it is assumed that the integral exists. We apply the idea here to the resonance situation in the electrical circuit of Chapter 15.

Referring back to Section 15.1, the governing differential equation is:

$$\frac{d^2 I}{dt^2} + \omega^2 I = \frac{\omega E_0}{L} \cos \omega t, \text{ with } \omega^2 = 1/LC.$$

Apply the transform to the entirety of this equation, so that:

$$\int_0^\infty e^{-st} \left[\frac{d^2 I}{dt^2} + \omega^2 I \right] dt = \frac{\omega E_0}{L} \int_0^\infty e^{-st} \cos \omega t \, dt,$$

---
[1] Johann Radon, 1887 to 1956.

and define the Laplace transform of $I(t)$ as $K(s)$. Therefore,

$$\int_0^\infty e^{-st}\frac{d^2I}{dt^2}dt + \omega^2 K(s) = \frac{\omega E_0}{L}\int_0^\infty e^{-st}\cos\omega t\, dt. \qquad (20.1)$$

We work out the last integral first, integrating by parts:

$$M = \int_0^\infty e^{-st}\cos\omega t\, dt = \left[\frac{1}{\omega}e^{-st}\sin\omega t\right]_0^\infty + \frac{s}{\omega}\int_0^\infty e^{-st}\sin\omega t\, dt,$$

$$= \frac{s}{\omega}\int_0^\infty e^{-st}\sin\omega t\, dt, \text{ after applying the integration limits,}$$

$$= \frac{s}{\omega}\left\{\left[-\frac{1}{\omega}e^{-st}\cos\omega t\right]_0^\infty - \frac{s}{\omega}\int_0^\infty e^{-st}\cos\omega t\, dt\right\},$$

$$= \frac{s}{\omega}\left\{\frac{1}{\omega} - \frac{s}{\omega}M\right\}, \text{ since we have gone round in a circle.}$$

Solving this for M:

$$M\left(=\int_0^\infty e^{-st}\cos\omega t\, dt\right) = \frac{s}{s^2+\omega^2}.$$

Next look at the first integral in equation (20.1), and again integrate by parts:

$$\int_0^\infty e^{-st}\frac{d^2I}{dt^2}dt = \left[e^{-st}\frac{dI}{dt}\right]_0^\infty + s\int_0^\infty e^{-st}\frac{dI}{dt}dt,$$

$$= s\int_0^\infty e^{-st}\frac{dI}{dt}dt, \text{ using the initial condition on } dI/dt,$$

$$= s\left\{\left[e^{-st}I\right]_0^\infty + s\int_0^\infty e^{-st}I\, dt\right\} = s^2 K(s),$$

after applying $I = 0$ at $t = 0$. It may be noted that in the above integrals, the evaluation at infinity is implicitly assumed to be zero, on the basis that none of the functions (including I) can increase faster than the exponential function decays.

So putting the various integrals back into equation (20.1), we get:

$$s^2 K(s) + \omega^2 K(s) = \frac{\omega E_0}{L}\frac{s}{s^2+\omega^2},$$

or,

$$K(s) = \frac{\omega E_0}{L}\frac{s}{(s^2+\omega^2)^2}. \qquad (20.2)$$

This is the function in s-space, and it needs to be mapped back to the time domain in order to obtain $I(t)$. The general expression for the inverse Laplace transform is in complex space, in which s is treated as a complex number, and the contour integration that is needed can get a bit involved [59]. I have, however, included an outline derivation of the inversion in Appendix H — if

20.2. RADON TRANSFORM

only for my own satisfaction. So here we will wimp out[2] and refer to a table of Laplace Transforms (*e.g.* [59]), and we find that

$$\int_0^\infty e^{-st} \frac{t}{2a} \sin at \, dt = \frac{s}{(s^2 + a^2)^2}, \quad \text{for constant } a,$$

from which we deduce that

$$I(t) = \frac{E_0 t}{2L} \sin \omega t,$$

consistent with the result in Section 15.1 (and rather more simply obtained).

The application of integral transforms to partial differential equations is deferred to Chapter 25.

20.2 Radon Transform

This section provides a brief discussion of another transformation technique, the Radon Transform, that is now widely used in medical imaging, among other areas, and two such examples are examined below.

20.2.1 Medical Imaging

Many of us will have come across, or have needed, a CAT (Computerised Axial Tomography) scan at one time or another. This operates by firing numerous very fine pencil beam X-rays into tissue, to build up a map of the internal three-dimensional structure. Each such pencil beam will pass through various forms of tissue — muscle, bone or whatever — and will be attenuated according to the various densities and distributions involved. By itself, the output of one such beam will not be particularly informative, but combining the results of numerous such beams from different angles, the internal structure can be determined.

Suppose that the tissue structure in a two dimensional slice is characterised by a function $f(x, y)$ in terms of Cartesian coordinates x and y. Then the Radon Transform is defined as [60]:

$$R(\rho, \theta) = \iint_{x,y=-\infty}^{\infty} f(x,y)\, \delta\left(x \cos\theta + y \sin\theta - \rho\right) dx\, dy, \tag{20.3}$$

which provides a mapping from the original coordinates x and y into a new pair, ρ and θ. The function $\delta(.)$ is the *Dirac delta function*[3], defining $\delta(x - x_0)$ — in very loose terms — as being infinite in magnitude if $x = x_0$ and zero everywhere else. Its effect in the integral in equation (20.3) will — for any ρ-θ pair — be to concentrate the entire result into those values of x and y that happen to be consistent with the equation

$$x \cos\theta + y \sin\theta - \rho = 0. \tag{20.4}$$

---

[2]Using such a table is actually normal practice; working in the complex space would generally be adopted only as a last resort (or if one really likes a challenge).
[3]Paul Dirac, 1902 to 1984.

This, however, is the general equation for a straight line in x-y space, so equation (20.3) is integrating $f(x,y)$ over all those values of x and y lying on that particular line[4].

Another way of looking at this mapping is to see that an entire line in x-y space is equivalent to a single point in ρ-θ space. A straight line is not the only possibility, of course: other geometric shapes inside the delta function — such as circles — are also possible [61], although we will stick with the linear case here.

As an aside, a useful symmetry relation for equation (20.3) is that $R(\rho,\theta) = R(-\rho,\theta+\pi)$.

We can now see how this transform applies to the CAT scan, since each pencil beam X-ray corresponds to one such straight line. By rotating the CAT machine through a circle, numerous such lines can be made to fill a disc, which will be one thin cross-section through the tissue. Many such slices may then be fitted together to form the scan data from which a full three-dimensional image may be retrieved.

To expand a bit on equation (20.4) for a straight line in the plane, refer to Figure 20.1.

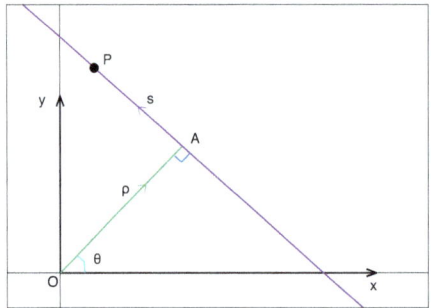

Figure 20.1: Radon Transform coordinate mapping

The function f is defined in terms of Cartesian x and y, while the two coordinates defining the line of interest (coloured purple) that satisfies equation (20.4) are ρ and θ. Point A is the nearest approach of the line to the origin O, so the line OA (coloured green) meets the purple line at a right angle, and A is a distance ρ from O. The line OA makes an angle θ to the horizontal x axis.

Any point P along the purple line can then be defined in terms of a distance s from A (which can be positive or negative), while point A is readily given by the vector $(\rho\cos\theta, \rho\sin\theta)$. So if P has coordinates x and y, then in vector terms

$$(x,y) = \rho(\cos\theta, \sin\theta) + s\,\hat{a},$$

---

[4]Equation (20.4) may not look to be in 'familiar form', but if it is divided throughout by $\sin\theta$ and y is then separated out on its own to the left, we do end up with the familiar $y = \alpha + \beta x$, α and β being functions of ρ and θ. Except that this form is invalid when $\sin\theta = 0$, whereas equation (20.4) defines a line for any θ.

20.2. RADON TRANSFORM

where $\hat{a}$ is the unit vector from A to P. But since $\hat{a}$ is perpendicular to the line OA, $\hat{a} = (-\sin\theta, \cos\theta)$, and so we end up with

$$x = \rho\cos\theta - s\sin\theta, \quad y = \rho\sin\theta + s\cos\theta.$$

In turn, these two relations mean that equation (20.3) for the Radon Transform can also be written in the form

$$R(\rho, \theta) = \int_s f(\rho\cos\theta - s\sin\theta, \rho\sin\theta + s\cos\theta)\,ds, \tag{20.5}$$

which just integrates along the purple line in Figure 20.1 using the path length parameter s between whatever limits are appropriate; this form of the integral is a bit easier to manage in numerical terms than is equation (20.3).

To extract the required image from the data recorded in $R(\rho,\theta)$ requires inverting the transform, and a filtered back-projection technique is generally used. This takes the deceptively formidable[5] form [61]:

$$f(x,y) = \int_{\theta=0}^{\pi} \int_{\nu=-\infty}^{\infty} |\nu| \left\{ \int_{\rho=-\infty}^{\infty} p_\theta(\rho)e^{2\pi i\rho\nu}\,d\rho \right\} e^{-2\pi i\nu(x\cos\theta + y\sin\theta)}\,d\nu\,d\theta, \tag{20.6}$$

in which $p_\theta(\rho)$ stands for $R(\rho,\theta)$ keeping θ fixed (which just highlights the requirement that the innermost integral in equation (20.6) is to be evaluated for each value of θ individually, before proceeding to the integrals over ν and then θ).

To add a bit more background information on equation (20.6), the innermost integral is a Fourier Transform from ρ-space to frequency ν-space; the quantity i is $\sqrt{-1}$ as usual. The second double integral over ν and θ is an inverse Fourier Transform in polar coordinates. Equation (20.6) is in reality a bit easier to derive than might be apparent and an outline of the process may be found in Appendix I.

You may have noticed that there is a 'missing link' in the above discussion: what is the CAT scanner actually measuring? During their passage through tissue, X-rays are attenuated to differing degrees depending on the density of that tissue, and the relationship between the source and collector intensities is given by the Lambert-Beer law [62]:

$$I = I_0 \exp\left\{-\int_s f(x(s), y(s))\,ds\right\}.$$

Here, I is the beam intensity after passing through the tissue, I_0 is the original intensity, s is the path length and $f(x(s), y(s))$ is the tissue density — a function of location. Taking the logarithm of this and reversing the signs on both sides, we get:

$$\log_e\left(\frac{I_0}{I}\right) = \int_s f(x(s), y(s))\,ds, \tag{20.7}$$

---

[5]You know you need to be worried if an estate agent describes your house as deceptively spacious ...

and it is this cumulative density on the right-hand-side that is measured in the above Radon Transform.

Presenting equation (20.6) is all very well, but it is understandable that there may be some lingering doubts that it can actually be used to extract the internal structure of a scanned sample. In the absence of recorded medical data, it is fortunately not too difficult to simulate a slice of tissue using equation (20.7) for some very simple nested density structures, as shown in the left-hand picture of Figure 20.2. The output from equation (20.6) is contained in the right-hand picture.

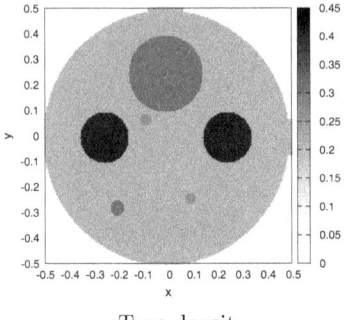 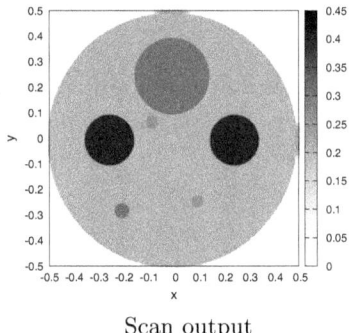

True density Scan output

Figure 20.2: Simulated CAT scan output

The simulated densities (with entirely arbitrary values) are shown on the scales to the right of each picture. Close examination shows a small amount of 'speckle' at the boundaries between the different density blocks[6], but the results are quite decent, despite an absence of fine-tuning in the discretised version of equation (20.6).

20.2.2 Straight Line Extraction

The Radon Transform also has application in areas other than medical imaging. Reference [61] is concerned with Synthetic Aperture Radar data processing, which I do not intend to discuss further. Another application is the extraction of straight lines in two-dimensional images, such as might be provided by a composite time series of oblique looks at a sea surface in which a faint object is intermittently seen. Much of the time the object will be obscured by waves, and it is only by means of integrating along its trajectory that there is a chance of extracting it from the background. In this particular context, the technique is more commonly known as the Hough[7] Transform.

This sort of situation is readily simulated by setting up a grid of pixels, and arranging for a hypothetical sensor to provide a measure of signal strength for each pixel. If the pixel corresponds to the known trajectory, the signal

---

[6]More visible when using colours, but the use of grey-scales seems to be standard medical practice.
[7]Paul Hough, 1925 to 2021.

20.2. RADON TRANSFORM

strength is set slightly higher than for the pixels containing background wave returns. Randomising the signal strength values then simulates the effects of intermittent viewing, wave obscuration, *etc.* For the present illustration, it is convenient to assume an exponential probability density (for details, see Section F), with mean strength values set to 1 for the waves and 2 for the object. Their respective probability density functions are shown in Figure 20.3.

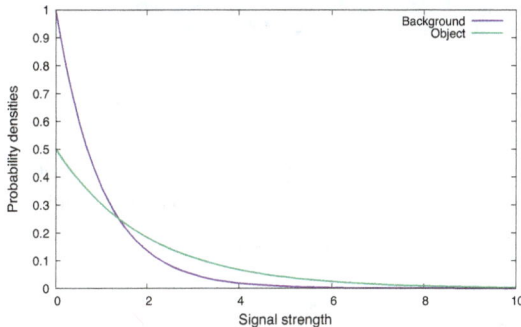

Figure 20.3: Probability density functions for background and object

Since the distributions overlap so much, it can be expected that the object will be correspondingly difficult to pick out, and this is supported by the results in Figure 20.4 — even after the lower signal strength values have been omitted from the picture. There is some faint evidence for a line heading diagonally from about bottom-left to top-right, but is it real? By integrating along the line, it should be possible to find out.

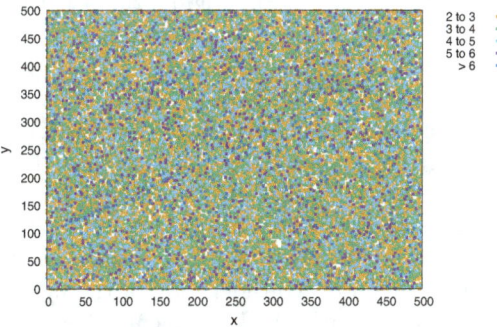

Figure 20.4: Signal strength values in x-y space, values > 2

Applying equation (20.5) for $0 \leq \rho \leq 500\sqrt{2}$ and $0 \leq \theta \leq \pi$, with small steps in each, and integrating in small increments in the path length s, we get the following picture in ρ-θ space.

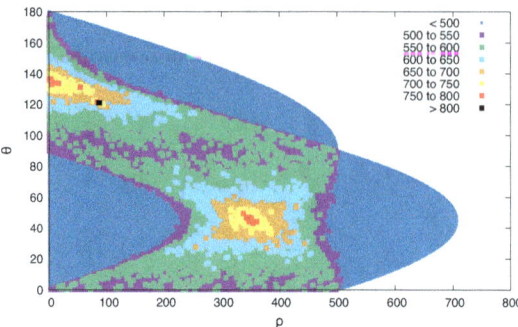

Figure 20.5: Integrated signal strength values in ρ-θ space, values > 500

The 'hot spots' with integrated strength above 750 are coloured red and black, and if these sets of points are extracted and turned back into lines, we get Figure 20.6.

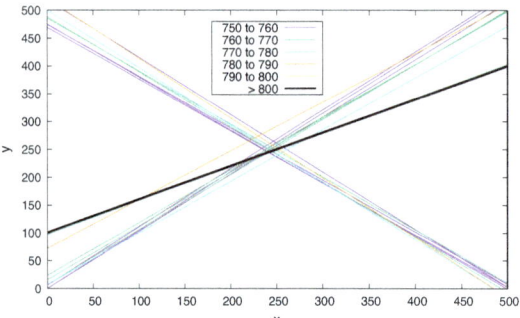

Figure 20.6: Candidate lines, integrated strength values > 750

The correct solution is the black line in Figure 20.6, which actually consists of two near-coincident lines, and comprises the two integrated values exceeding 800 marked by the black squares in Figure 20.5.

The other lines retained in Figure 20.6 — and all of the others with lower integrated strength values which have been omitted — occur 'at random'; the cumulative signal strength is enough to exceed whatever threshold has been imposed. A way to obviate these would be to apply a filter to the raw signal strength values, although with the drawback that the resulting data will be that much sparser. Setting suitable filter thresholds, either on the raw data or on the integrated signal strength, requires consideration of the underlying probability distributions and is rather beyond what I can include here [63].

Chapter 21

KALMAN FILTERS

THIS is, perhaps, an unusual topic to find here, it being more usually discussed in books on engineering (and rather specialised books at that, for example [64]). The term 'filter' suggests noise reduction, which is indeed one aspect of the Kalman[1] Filter, the other being the ability to extract additional information from data streams.

Reduced to its essentials, the Kalman Filter fuses two independent sources of information: a set of measurements, and a mathematical model of how the object being measured is expected to evolve in time. This fusion process acts to produce a result that is more than the sum of its parts: typically reduced uncertainty and the extraction of information that is not present in the measurements themselves.

It is simplest to use an example, namely tracking a sequence of radar measurements. In this context, a 'track' consists of time, position, velocity, perhaps acceleration and potentially any other attributes of the object in question that can be measured or inferred, plus associated uncertainties. The radar provides only discrete position measurements, each with a time stamp; successive measurements on the same object can be fractions of a second apart, but are more usually of the order of several seconds, depending on the radar. In between these measurements, the track needs to be propagated in time, which is where the mathematical model comes in. So the track from an earlier time is propagated to the time of the most recent measurement, and this propagated track is then fused with that measurement. The result will be a refined track at the most recent time — refined in the sense that uncertainties will have been reduced (or at least be no larger than before). An additional benefit of the mathematical model is that velocity, acceleration and so forth (depending on the model) are automatically obtained from a sequence of measurements — a refined version of 'join the dots', if you will.

The systematic treatment of uncertainty is of central importance, and at this point it is worth making the distinction between accuracy and uncertainty. Put simply, accuracy is a measure of the actual error — how far off the track is from the real position of the object (the 'ground truth'). Uncertainty is our

---

[1] Rudolph Kalman, 1930 to 2016.

estimate of that error. Ideally, uncertainty should be a decent reflection of the error magnitude, but this need not be the case — the track could be wonderfully certain, but in completely the wrong place.

Section 4.2 introduced the *standard deviation* as a measure of uncertainty for a single quantity, the *variance* being the square of the standard deviation. We now need to deal with *error covariance matrices*, which are a straightforward extension: each element of such a matrix (see Chapter 14 for some introductory stuff on both vectors and matrices) contains the variance for that particular state component, or the cross-component for a mix of states. Continuing with the above example, suppose the state contains only position and velocity in one dimension, written in the vector form:

$$\mathbf{x} = (x, u),$$

where u is the velocity component, standing for the time derivative of x. The covariance matrix P encapsulating the uncertainty associated with $\mathbf{x}$ is then the 2×2 matrix

$$P = \begin{pmatrix} P_{xx} & P_{xu} \\ P_{ux} & P_{uu} \end{pmatrix}.$$

The components are more handily written as P_{ij}, with the i and j running over 1 to 2, and using the convention that $i = 1$ corresponds to x and $i = 2$ to u.

A feature of covariance matrices are that they are symmetric: $P_{ij} = P_{ji}$, which reduces the number of independent quantities (in this case from 4 to 3).

21.1 Kalman Filter Equations

Now for the discrete Kalman Filter vector-matrix equations (the notation is basically that of [64]):

Stage 1: propagation from time t_k to time t_{k+1}:

$$\mathbf{x}_{k+1}(-) = \Phi \mathbf{x}_k + \mathbf{c}, \tag{21.1}$$

$$P_{k+1}(-) = \Phi P_k \Phi^T + Q. \tag{21.2}$$

The matrix Φ encapsulates the expected motion of the object, which may be as simple as straight-line dynamics, while matrix Q describes that model's perceived deficiencies. Small Q implies high confidence and vice versa. Matrix P describes the uncertainty associated with the state $\mathbf{x}$, while the superscript T denotes matrix transpose — essentially twiddling it about its main diagonal line stretching from top-left to bottom-right. The subscript k just acts as a time update counter.

Vector $\mathbf{c}$ is independent of $\mathbf{x}$ and can act as a control; its inclusion will become a bit clearer in Section 21.6 below.

Stage 2: calculation of the Kalman Gain matrix K:

$$S = HP_{k+1}(-)H^T + R, \qquad (21.3)$$
$$K = P_{k+1}(-)H^T S^{-1}. \qquad (21.4)$$

Matrix R defines the measurement uncertainty, while H describes the relationship between the measurement and the state vector.

Stage 3: incorporation of the sensor measurement z:

$$\mathbf{x}_{k+1}(+) = \mathbf{x}_{k+1}(-) + K\left[\mathbf{z}_{k+1} - H\mathbf{x}_{k+1}(-)\right], \qquad (21.5)$$
$$P_{k+1}(+) = (I - KH) P_{k+1}(-). \qquad (21.6)$$

H can reflect the fact that the radar may only be able to measure position and not velocity, so the measurement **z** is smaller (in number of dimensions) than is **x**. I is the unit matrix — unity on the main diagonal and zero elsewhere — while the $(-)$ and $(+)$ notations are used simply to mark pre- and post-incorporation of the sensor data at the same time stamp.

The above set of six recursive equations are readily solved on a computer, with the output from stage 3 being fed back to the start of stage 1. But as they stand, it is not obvious what the various quantities actually *do*, and this can be explained by means of a pair of simple examples.

21.2 Estimation of a Constant

This is about the simplest example possible — estimation of a constant, given a sequence of noisy measurements (there are links here back to Section 4.2). The above matrices then become scalar quantities, with $\Phi = H = 1$, in which case we have the following simplified set:

$$P_{k+1}(-) = P_k + Q,$$
$$S = P_{k+1}(-) + R,$$
$$K = \frac{P_{k+1}(-)}{S} = \frac{P_{k+1}(-)}{P_{k+1}(-) + R}, \text{ using the expression for } S.$$

Therefore, factoring in the first of the above equations, we get for the Kalman Gain:

$$K = \frac{P_k + Q}{P_k + Q + R}.$$

This is just a ratio of uncertainties: the top bit describing the prior uncertainty plus the uncertainty associated with the mathematical model, while the bottom bit factors in the measurement uncertainty R. Since all of the quantities involved should be positive, we can see that $0 \leq K \leq 1$ in this one-dimensional example.

Now have a look at how K affects the updated state x_{k+1}:

$$x_{k+1}(+) = x_{k+1}(-) + K\left[z_{k+1} - x_{k+1}(-)\right].$$

For small R relative to $P_k + Q$, K will be close to 1, in which case $x_{k+1}(+) \approx z_{k+1}$. That is, accurate sensor data means that the updated x closely follows the measurements. Alternatively, if R is large relative to P_k+Q, K will be small and the updated x remains closer to the mathematical model. Both aspects are what we would expect intuitively.

Q reflects our estimate of how good the model is — small Q implying high confidence and larger Q implying less confidence. So $Q = 0$ means that we are *really* certain that what we are measuring is genuinely constant, in which case

$$K = \frac{P_k}{P_k + R},$$

and so, incorporating K into the updated error covariance P,

$$P_{k+1} = \frac{RP_k}{P_k + R}.$$

Since this is independent of randomness in the measurements themselves, it is possible to work out how P_k varies as a function of k, on the assumption that R is a constant. So turn this equation upside down:

$$\frac{1}{P_{k+1}} = \frac{1}{P_k} + \frac{1}{R},$$

which has the solution

$$\frac{1}{P_k} = \frac{1}{P_0} + \frac{k}{R}.$$

(To see this, start from $k = 0$ and work out $1/P_1$, and then $1/P_2$, and so on). The quantity P_0 represents the initial uncertainty on our estimate of the constant we are measuring, so in the absence of any other information we may as well set $P_0 = R$, in which case

$$P_k = \frac{R}{k+1}.$$

So for large k, the standard deviation associated with the updated x decreases as $1/\sqrt{k}$, as stated in Section 4.2.

21.3 Fusion of Radiometric Dates

An interesting variant on the preceding example is also concerned with obtaining a better estimate of some constant value, but where the various data samples are uncertain to differing extents. Borrowing some figures from [12], independent estimates of the age of the Permo-Triassic boundary are stated to be 250 ± 6

21.3. FUSION OF RADIOMETRIC DATES

million years BP (before present), 251.1±3.4 MY BP and 249.9±1.5 MY BP. As a reminder, the end-Permian event is estimated to have eliminated somewhere between 90% and 95% of all species around at that time — sufficiently traumatic to have captured the attention of many palaeontologists and geologists[2].

Looking at the above three sets of Permo-Triassic boundary dates, it is tempting to concentrate entirely on the third, since this is quite obviously more certain than the other two. But this would be to ignore relevant data; the situation is not unlike that discussed in Section 6.2 above, in which rather more definitive answers can be obtained if we consider *all* of the relevant information/data. Indeed, there are strong links between the largely statistical fusion methods that are examined in the present chapter and the fusion of probabilistic values (Bayes' theorem) illustrated in Section 6.2.

The question then is, can we obtain an even more certain date estimate by fusing all three of the above values? The results from Section 21.2 certainly implies so, and it will be demonstrated below that this is the case.

We use a similar notation to that in the previous section: z stands for a measurement, σ for its standard deviation and R for its variance; x for a fused value and P for its variance (the square of the associated standard deviation). So we have, as source data:

$$z_1 = 250, \sigma_1 = 6, R_1 = \sigma_1^2 = 36,$$
$$z_2 = 251.1, \sigma_2 = 3.4, R_2 = \sigma_2^2 = 11.56,$$
$$z_3 = 249.9, \sigma_3 = 1.5, R_3 = \sigma_3^2 = 2.25.$$

The standard deviations here have been taken directly from the above ± values, which are stated to be assessments of experimental error. Correctly, the σ values would be derived having some knowledge of the underlying error distribution, but the distinction is not especially important for the present example, where it is the process that is relevant rather than the final answer.

Start with the first measurement. In the absence of any other information (such as preceding date estimates), we set $x_1 = z_1$ and $P_1 = R_1$; the subscripts here act as convenient counters for the fusion stages.

Now move on to the second measurement. In common with the assumptions of preceding section, we set $\Phi = 1$ and $Q = 0$, so that the associated Kalman Gain at this stage is:

$$K = \frac{P_1}{P_1 + R_2}.$$

(We don't need a subscript here, since K is a temporary quantity). Therefore,

---

[2]The trigger of this event still remains uncertain, but massive and prolonged volcanic eruptions is a viable candidate. There is no unequivocal evidence for an asteroid impact at that time, in contrast to the K-Pg boundary date mentioned in Chapter 5.

the second fused date estimate will be given by:

$$x_2 = x_1 + K(z_2 - x_1),$$
$$= x_1 + \frac{P_1}{P_1 + R_2}(z_2 - x_1),$$
$$= z_1 + \frac{R_1}{R_1 + R_2}(z_2 - z_1), \text{ using } x_1 = z_1 \text{ and } P_1 = R_1 \text{ from above,}$$
$$= \frac{R_2 z_1 + R_1 z_2}{R_1 + R_2} \text{ after simplifying the algebra.}$$

This is just the weighted average of the two individual data values. In a similar manner, the associated variance for x_2 is given by:

$$P_2 = (1 - K) P_1,$$
$$= \left(1 - \frac{P_1}{P_1 + R_2}\right) P_1 = \left(1 - \frac{R_1}{R_1 + R_2}\right) R_1,$$
$$= \frac{R_1 R_2}{R_1 + R_2}.$$

Fusing in the third measurement z_3 follows exactly the same logic. It is instructive to follow through the algebra and derive specific analytic results, although in practice the recursive equations would be used as part of a numerical derivation scheme. Omitting the algebra, we get:

$$x_3 = \frac{R_2 R_3 z_1 + R_1 R_3 z_2 + R_1 R_2 z_3}{R_1 R_2 + R_1 R_3 + R_2 R_3}, \text{ (also a weighted average),}$$
$$P_3 = \frac{R_1 R_2 R_3}{R_1 R_2 + R_1 R_3 + R_2 R_3}.$$

There are patterns here, but continuing with the above explicit process is going to become rather cumbersome (quite apart from chewing up paper). In any case, there are now no more measurements to include, so we can stop, plug in the numeric values provided above and see what we get. After a bit of punting on a calculator, the numbers become:

$$x_3 = 250.0907 \text{ and } \sigma_3 = 1.3378.$$

This fused estimate (coloured green) is graphed in conjunction with the source data (in purple) in Figure 21.1.

21.4. CONSTANT VELOCITY

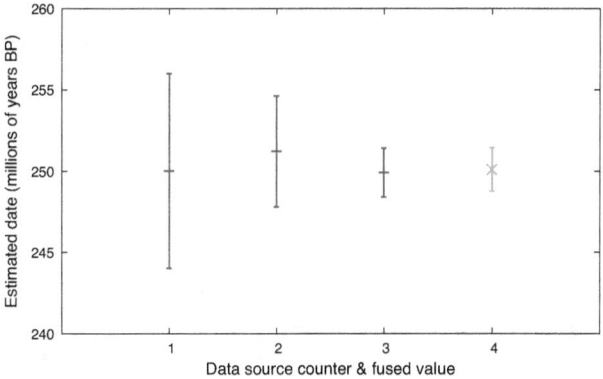

Figure 21.1: Radiometric dates, uncertainties and fused estimate

There is obviously a slight change from z_3 to x_3, accompanied by a small (but potentially valuable) reduction in the uncertainty associated with x_3 when compared to σ_3. This might have been expected, since the third date value was so much more certain than the other two, but the Kalman Filter has fulfilled its promise of reducing uncertainty (or at least not making it any larger).

All of the source data points shown in Figure 21.1 are consistent, in that they overlap with no outliers. It may be the case, though, that one or more of them are visually (and potentially statistically) incompatible with the others. For instance, z_2 could come out as 310 MY BP with the same associated standard deviation, perhaps due to sample contamination or whatever. Statistical tests based on the degree of overlap in probability distributions provide a measure of confidence — typically a probability of match or no match — when accepting or rejecting data points.

Data fusion, either using the statistically-based approach described here or the probabilistic form of Section 6.2, constitutes what is termed *meta-analyses*. For example, the results of several small medical trials into the efficacy of some treatment or other can be pulled together in a systematic manner to improve the overall level of confidence. In a very real sense, the whole is greater than the sum of the parts.

21.4 Constant Velocity

A third simple example helps understand how we can get to the Kalman Filter equations from kinematic equations of differential form — a fairly common occurrence. Suppose our mathematical model is the second-order differential equation:

$$\frac{d^2x}{dt^2} = \nu,$$

where t is time and ν stands for random disturbances (of which more below).

Cast this into a pair of first-order equations:

$$\frac{dx}{dt} = u,$$
$$\frac{du}{dt} = \nu.$$

(To retrieve the original second-order equation, differentiate the first of these with respect to time and substitute the second).

The discrete Kalman Filter expects measurements at distinct time intervals, so these differential equations need to be replaced by a pair of difference equations (Section 4.1 introduced difference equations in another context). Let T be the time step between measurements, assumed small for the time being; then:

$$\frac{dx}{dt} \approx \frac{x_{k+1} - x_k}{T} = u_k,$$
$$\frac{du}{dt} \approx \frac{u_{k+1} - u_k}{T} = \nu.$$

using index k to denote time step. Rewrite these in the form:

$$x_{k+1} \approx x_k + T u_k,$$
$$u_{k+1} \approx u_k + T \nu.$$

Define the vector $\mathbf{x}_k$ in column form as the pair $(x_k, u_k)^T$ and write the difference equations in vector-matrix form:

$$\mathbf{x}_{k+1} = \begin{pmatrix} 1 & T \\ 0 & 1 \end{pmatrix} \mathbf{x}_k + \begin{pmatrix} 0 \\ T\nu \end{pmatrix},$$
$$= \Phi \mathbf{x}_k + \mathbf{v},$$

thus defining the transition matrix Φ. It can be seen that the random bit present in the quantity ν has been separated out, so assume that ν has a mean value of zero and standard deviation s. Therefore vector $\mathbf{v}$ has zero mean and the uncertainty inherent in the above model is encapsulated in the matrix Q, with

$$Q = \begin{pmatrix} 0 & 0 \\ 0 & T^2 s^2 \end{pmatrix}.$$

This is a simple example of the matrix Q that appears in equation (21.2) for the uncertainty propagation.

It may be remarked that the randomness affecting the above difference equations is assumed at a constant value over the time interval T, with a different random value for the next interval. Other types of random disturbances are also possible, one of the more common ones being 'white noise': of infinitesimal magnitude but occurring infinitely often over the time interval [64]. How to model random disturbances needs to be guided by the physics of the situation.

21.5 Propagation of Uncertainty

Look back now at equation (21.2) for the propagation of the error covariance matrix in time. An unstated assumption here is that the kinematic model is *linear*, meaning that the transition matrix Φ is at most a function of time (and not a function of **x** itself). So what to do when we are dealing with earth gravity which — from Section 11.4 — is most definitely nonlinear? This question has relevance when working out options for intercepting a ballistic missile with another (interceptor) missile: the current location and velocity of the threat will be uncertain, with this uncertainty characterised in matrix P, so how do we evolve P forward[3] in time?

Without going into over-much detail, several choices are available here:

- Approximate the nonlinearity with linearised equations. This can be made to work adequately if the propagation time intervals are not too long, but is otherwise unsatisfactory.

- Recognise that the state vector **x** and covariance P are actually statistical characterisations of a probability distribution (**x** and P are, essentially, mean and standard deviation squared). This suggests that instead of working with such statistical measures, as in the Kalman Filter, we sample the distribution: find a random set of points within that distribution which can then be propagated nonlinearly. Each such point represents a candidate position and velocity, and can be propagated nonlinearly as far in time as needed without compromise.

  This approach forms part of what is known as a *particle filter* [65].

- The above particle filter, while systematic, can nonetheless require very many particles and be costly in computational terms. A simpler option that trades a bit of accuracy for computational efficiency uses a much smaller set of (typically symmetrically distributed) particles, each of which can be propagated nonlinearly as before [66].

In both of the second two options, the location and uncertainty are characterised by a cloud of points. In general, the effect of propagation in time is to make the cloud larger, so the uncertainty increases. This has obvious relevance to the missile intercept problem: many interceptors have their own seekers, whether radar or optical, so ground control needs to get the interceptor into a position where it can 'see' the threat for itself. If the uncertainty is too large, the interceptor may well not find it.

Continuing with the recognition that what we are actually dealing with is a probability distribution (whether liking it or not), it is worth writing down the partial differential equation governing the evolution of such an animal. For one spatial dimension x and time t, the Fokker-Planck equation [67] is:

$$\frac{\partial}{\partial t} p(x,t) + \frac{\partial}{\partial x}\left(v(x)p(x,t)\right) = \frac{\partial^2}{\partial x^2}\left(D(x)p(x,t)\right)$$

---

[3] Or backward, for that matter, to estimate the threat launch point.

where $p(x,t)$ is the distribution. This is a diffusion-type equation, the left-hand-side describing transport and the right-hand-side characterising dispersion, with similarities to the flow of fluids; but that is as far as I intend to go in explaining any further.

Right at the start of this chapter, the term 'fusion' was used in the introductory words on the Kalman Filter. Although not obvious, there are strong links between data fusion in the tracking context, as above, and in Bayes' rule (Chapter 6): both are concerned with the systematic inclusion of new information. The link is a bit more evident with particle filters, which make use of the Chapman-Kolmogorov equation [68], which I cannot resist including (although again without further explanation):

$$p(x_1, t_1 | x_3, t_3) = \int p(x_1, t_1 | x_2, t_2) p(x_2, t_2 | x_3, t_3) dx_2.$$

21.6 Missile Intercept

The inclusion of this section of the book was prompted by a comment from a friend, remarking on the complexities involved in detecting the launch of a ballistic missile, working out where it is going, and then deciding whether an interceptor missile can be sent out to destroy the threat before it lands. It is not intended here to go into the specifics of any particular threat missile or interceptor type, nor to discuss the more detailed engagement strategies. Rather, the aim is to use the intercept problem as a convenient forum for introducing the Kalman Filter equations and highlighting the central rôle of uncertainty.

We use the missile simulation in Figure 21.2 as a basis. Motion is from left to right, while the radar and interceptor launch point (here coincident) are marked by the blue square on the right. Distances are in kilometres.

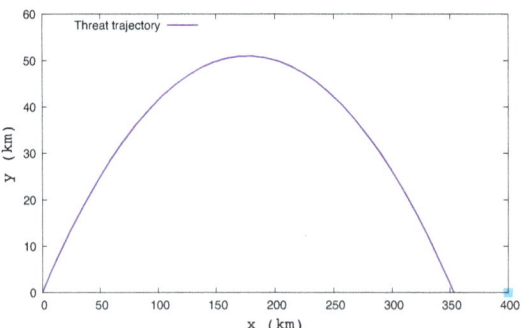

Figure 21.2: Example threat trajectory

To simplify the tracking logic, two-dimensional motion over a flat earth is assumed, with constant gravitational acceleration and no air resistance (see Section 11.1). The radar provides measurements at one-second intervals, with 50 metre uncertainties in both x and y. These simplifications enable the use of a linear Kalman Filter, as listed in Section 21.1 above; correctly, the gravitational

21.6. MISSILE INTERCEPT

model would be nonlinear, and the radar would provide range and angle instead of x and y.

To go any further, it helps to outline how we can design an intercept logic for the above situation. Suppose that the missile has been tracked at a time t_0, meaning that estimates of position, velocity and their associated uncertainties are available at that time. The interceptor requires a finite time to fly out, so it is necessary to predict the missile track ahead by some interval T to a point $\mathbf{x}$, and then ask the question, "can the interceptor reach this point in the time available?". In mathematical terms, if $\mathbf{x}$ denotes the missile position and $\mathbf{s}$ that of the interceptor, then

$$\mathbf{x}(t_0 + T) \approx \mathbf{s}(t_0 + T),$$

the approximation sign denoting the fact that it is only necessary to get the interceptor near enough to the missile for it to be able to 'see' it and then carry out the final engagement autonomously.

In the simplified logic used here, the interceptor is assumed to be launched at the same track time t_0. This is not strictly necessary, as an interceptor launch time $t_1 \neq t_0$ could be used, in which case we would need $\mathbf{x}(t_0+T) \approx \mathbf{s}(t_1+\tau)$ for some other time interval τ. This is an unnecessary complication for the present illustration, though.

Another simplification can also be made here: that the interceptor uses a straight line trajectory, which minimises its flyout time. It is also assumed that the interceptor can travel at a fixed speed of V, in which case its flight time from launch to intercept is given by λ/V, where λ is the distance from interceptor launch point (the blue box in Figure 21.2) to the point $\mathbf{x}(t_0 + T)$.

But this is not the end of the matter: we also impose a condition on the extrapolated track uncertainty, that at t_0+T the position uncertainty should be small enough that the interceptor can find it on its own (which will be largely determined by the capabilities of its on-board seeker). This condition means, typically, that the threat missile needs to be 'in track' for some time before its extrapolated track uncertainty falls below the required threshold.

Putting the various bits together, we get the following logical sequence:

1. Use the Kalman Filter equations in Section 21.1 to obtain a track — position, velocity and associated uncertainties — at a sequence of times t_k. The transition matrix Φ is given by

$$\Phi = \begin{pmatrix} 1 & 0 & \Delta t & 0 \\ 0 & 1 & 0 & \Delta t \\ 0 & 0 & 1 & 0 \\ 0 & 0 & 0 & 1 \end{pmatrix},$$

where the state vector component ordering is $(x, y, \dot{x}, \dot{y})$ and Δt is the time interval between radar measurements. To incorporate gravity, it is necessary to define the following column vector $\mathbf{c}$ for use in equation (21.1):

$$\mathbf{c} = \left(0, -\frac{1}{2}g\Delta t^2, 0, -g\Delta t\right)^T,$$

This can be obtained from the basic differential equations $d^2x/dt^2 = 0$, $d^2y/dt^2 = -g$, and linearising over the interval t_k to t_{k+1}.

2. Extrapolate this track forward incrementally in small time steps τ, using equations (21.1) and (21.2). In the present case, τ can be the same as Δt, but one would normally make it smaller to cope with nonlinearities and to provide a finer-grained solution.

3. At each incremental time $t_k + n\tau$, with integer n and $n\tau = T$, check if the following conditions hold:

    - $\sqrt{\frac{1}{2}(P_{xx} + P_{yy})} \leq \chi$, for some uncertainty threshold χ. This basically says that the net extrapolated missile track position uncertainty is now small enough for the interceptor to operate autonomously.
    - $\frac{\lambda}{V} \leq T$. This condition requires that the interceptor has enough speed to reach the extrapolated missile position in the time available, assuming it (the interceptor) is launched at time t_k. Correctly, there would also be a maximum interceptor range governed by the amount of on-board fuel, but this is ignored here.

If both conditions are satisfied, a candidate intercept solution is obtained. Otherwise, return to step 1 and continue tracking.

Figure 21.3 provides an illustration of the earliest such solution, using $\chi = 1$ kilometre and measuring times from missile launch. The missile track is extrapolated ahead from a time of 16 seconds to 105 seconds, which the interceptor can also reach in the same time interval of 89 seconds if launched at $t = 16$ seconds. The interceptor speed has here been assumed to be 2.5 kilometres per second, a bit faster than the missile itself, which was launched at a speed of 2 kilometres per second. The interceptor missile here (its path being the black line) heads directly toward the predicted intercept point.

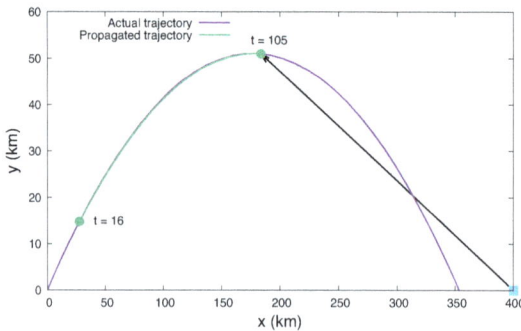

Figure 21.3: Intercept solution

The growth in the net missile position uncertainty over the same time interval is essentially linear, reaching 870 metres at 105 seconds, which is within the 1 kilometre limit.

21.7 Intercept Guidance Laws

The above simple example has concentrated on getting the interceptor to a point where it can 'open its eyes', so to speak, and then do the rest of the job by itself. It might be thought that heading directly toward the missile at each time value and at maximum speed would be adequate for a collision (or near-collision), but this is not necessarily the case, even if the target missile maintains the same course throughout. Even aiming at some point ahead of the missile need not work satisfactorily.

A more effective guidance law is known as 'proportional navigation', in which the intent is to adjust the interceptor acceleration such that the line of sight to the target remains constant. The difference between the two approaches is illustrated in Figure 21.4, here using an aircraft-type simulation with longer flight times. We'll come back to the ballistic missile example that is shown above in Figure 21.3.

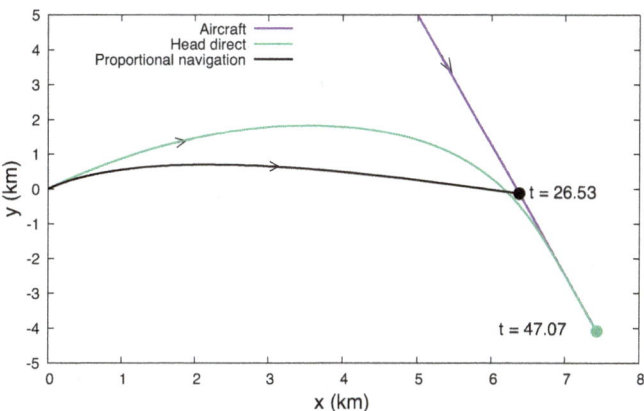

Figure 21.4: Simulated intercept guidance laws, aircraft case

Times are measured from interceptor launch.

Using vector $\mathbf{x}$ for the target location (in this case, an oncoming aircraft) and $\mathbf{s}$ for the interceptor, the equation for the 'head direct' option is:

$$\frac{d\mathbf{s}}{dt} = V\hat{\mathbf{a}}, \tag{21.7}$$

where V is the interceptor speed and $\hat{\mathbf{a}}$ is the instantaneous bearing vector between the aircraft and interceptor. That is,

$$\hat{\mathbf{a}} = \frac{\mathbf{x} - \mathbf{s}}{|\mathbf{x} - \mathbf{s}|};$$

(see Section 14). It can be seen that the interceptor ends up in a 'tail chase' situation, and could only succeed in this example because its speed was faster than that of the aircraft (250 versus 200 metres per second).

In its simplest form, proportional navigation (PN) uses the acceleration equation:

$$\frac{d^2\mathbf{s}}{dt^2} = \kappa V \frac{d\hat{\mathbf{a}}}{dt}, \tag{21.8}$$

where κ is a constant; in the present simulation the speed has been fixed at V and $\kappa = 6$.

Looking at Figure 21.4, we can see that the PN intercept time is only just over half that needed for the 'head direct option', and that for much of the interceptor's trajectory the line of sight is unchanging and the acceleration is near-zero. A zero acceleration also gives rise to a linear interceptor trajectory, which provides some justification for the corresponding assumption in Section 21.6.

The equivalent results for the simulated ballistic missile example from the preceding section are shown in Figure 21.5. It is here assumed that the intercept solution from Section 21.6 has been used to guide the interceptor to within 10 kilometres of the expected impact point, which corresponds to 4 seconds prior to the expected impact time of 105 seconds[4]. Thus Figure 21.5 covers only the time span 101 to 105 seconds, corresponding to the autonomous intercept 'end-game' at missile apogee in Figure 21.3.

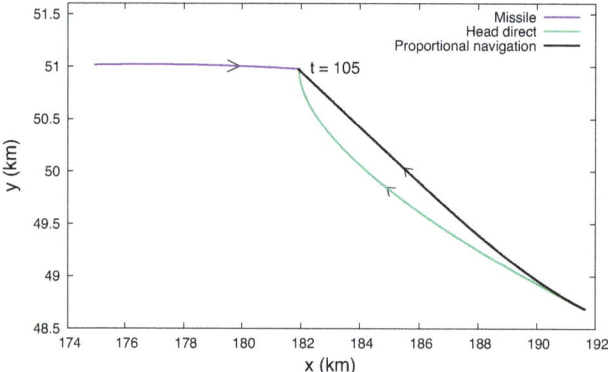

Figure 21.5: Simulated intercept guidance laws, ballistic missile case

There is still a difference between the two guidance laws, with proportional navigation giving a more direct trajectory. But the time scales are too short for there to be much of a difference in the impact times: proportional navigation ends at 105.017 seconds and 'head direct' at 105.063 seconds.

---

[4]The interceptor speed is here 2.5 kilometres per second.

Chapter 22

RANDOM WALK

THE concept of a random walk is invariably illustrated using a severely inebriated person, who commences at some position and then makes random steps in various arbitrary directions in a two-dimensional space. Quite why a drunkard is used for this purpose is unclear, since the following question was originally posed in the context of mosquito travel by Ronald Ross[1]: if a mosquito starts in the middle of a swamp and moves in random directions in search of a blood meal, how far will it have moved from the starting point in n steps, say? Such a question is evidently relevant from a medical or public health point of view.

Simulating such behaviour is straightforward enough, and Figure 22.1 is typical of what is found.

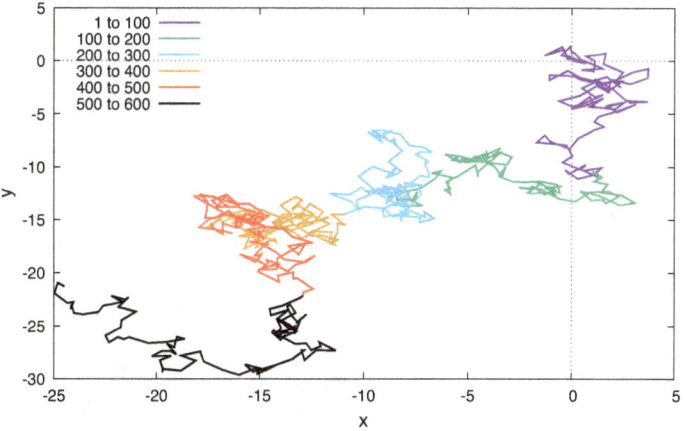

Figure 22.1: Example random walk

In this case, 600 steps have been simulated, each step being sampled from a zero-mean uniform distribution with a standard deviation of 0.6 metres[2]. The

---

[1] Ronald Ross, 1857 to 1932. He discovered the link between mosquitoes and malaria.
[2] I do not intend to carry out actual measurements to see if this value is appropriate. This

two Cartesian coordinates in x and y are generated in the same manner and are statistically independent. Blocks of steps are coloured differently to highlight the progress of the mosquito from its starting point at the origin.

Let us now have a look at what the mathematics predicts will happen, concentrating on the x coordinate since y will be similar. In discrete form, we have the difference equation

$$x_{k+1} = x_k + n_{k+1},$$

for step k, starting at the origin (so that $x_0 = 0$) and with the n_k being zero mean and having standard deviation σ. The following sequence will then be obtained:

$$x_1 = n_1,$$
$$x_2 = x_1 + n_2 = n_1 + n_2,$$
$$x_3 = x_2 + n_3 = n_1 + n_2 + n_3,$$
$$etc\ldots,$$

and $x_n = \sum_{j=1}^{n} n_j$ for n steps.

Now introduce a shorthand notation that will come in handy later on: the expectation operator E, such that for some function f,

$$E\{f(t)\} = \int_{-\infty}^{\infty} f(t)\, p(t)\, dt,$$

in which t stands for a continuous random variable underpinning each n_k and $p(t)$ defines the probability distribution associated with n_k. (In Figure 22.1, $p(t)$ is just the uniform distribution: namely, $1/(2\sigma\sqrt{3})$ if $-\sigma\sqrt{3} \le t \le \sigma\sqrt{3}$ and zero elsewhere).

Apply this operation to x_n:

$$E\{x_n\} = \int_{-\infty}^{\infty} \sum_{j=1}^{n} n_j(t)\, p(t)\, dt = \sum_{j=1}^{n} \int_{-\infty}^{\infty} n_j(t)\, p(t)\, dt,$$

$$= 0, \text{ since all of the } n_j \text{ are zero-mean.}$$

So, despite all of the the wandering back and forth and up and down, the mosquito is expected to remain in the vicinity of its starting point. This is the best that can be done in terms of its state estimate — the mean value, if you like. The *uncertainty* associated with that estimate, though, grows with the number of steps taken; it is a more useful quantity in working out the probability of finding the mosquito at some distance from the origin after n steps. This uncertainty can be determined simply from

$$E\{x_n^2\} = E\left\{\sum_{i=1}^{n} n_i \sum_{j=1}^{n} n_j\right\} = E\left\{\sum_{i=1}^{n}\sum_{j=1}^{n} n_i n_j\right\} = \sum_{i=1}^{n}\sum_{j=1}^{n} E\{n_i n_j\}.$$

---

is an illustrative simulation, after all.

It is now assumed that each random step taken is independent of its predecessors, which means that $E\{n_i n_j\} = 0$ unless $i = j$. So,

$$E\{x_n^2\} = \sum_{i=1}^{n} E\{n_i^2\},$$
$$= n\sigma^2, \qquad (22.1)$$

since the n_i are identically distributed.

What this means is that the uncertainty (standard deviation) associated with the mean value of x_n is expected to grow as $\sigma\sqrt{n}$.

"Hold on a minute", you may say, "it's not evident from Figure 22.1 that the mosquito stays near the start, nor that the uncertainty grows as you say it should". But Figure 22.1 shows just one out of the infinite set of possible trajectories the mozzie could take, since at each step there are infinitely many options. To illustrate a wider set of candidate motions, Figure 22.2 shows 20 such possibilities, with the colour coding now different for each candidate trajectory.

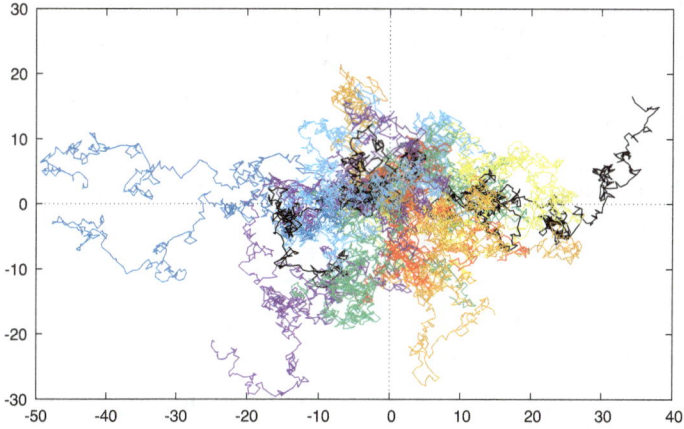

Figure 22.2: Example set of 20 random walks

It is now a bit more obvious that the tendency is to stay near the origin, although with a widening zone of uncertainty. Although numerically simulating an infinite number of trajectories is impractical, it is possible to see what happens if we generate a swarm of 5000 mosquitoes. Such a swarm can represent either 5000 possibilities for the same mozzie, or 5000 mosquitoes behaving in an identical (and independent) manner. With this more statistically meaningful sample, we can compute the mean location of the swarm at each time step, as well as its dispersion. In the former case, after 600 steps, the mean position was found to be $x = 0.057$, $y = -0.124$, which is consistent with staying put. The growth of the variances with the number of steps is shown in Figure 22.3, with the black line standing for equation (22.1); theory and simulation are in satisfactory agreement.

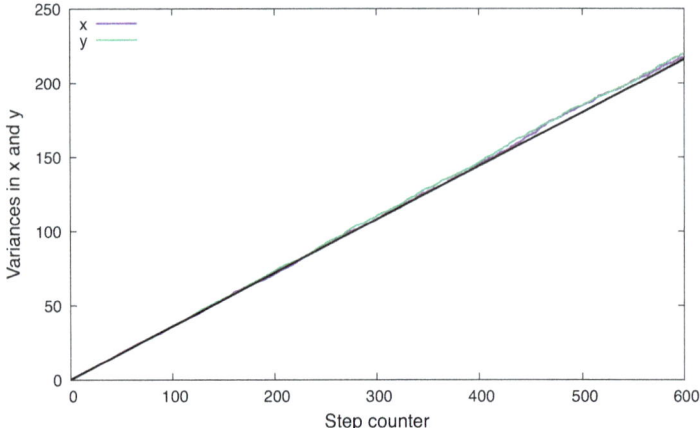

Figure 22.3: Random walk variances in x and y

The above conclusions may appear to be unduly vague — surely we should be able to do better! Unfortunately, no; in the absence of actual measurements of a particular mosquito's progress, we are stuck with the pure extrapolation of state and covariance. This should, perhaps, be no great surprise — after all, if a person goes missing, say, the search logically commences where that person was last located, with the search domain then expanding over time.

Indeed, casting the above problem into Kalman Filter terms provides a different view of the same situation and offers a different solution method, although we can only make use of Stage 1 in Section 21.1. In the present case, we have a scalar quantity x, for which $\Phi = 1$, $\mathbf{c} = 0$ and $Q = \sigma^2$. As in the above analysis, the mean value of the state x_n is zero, while the covariance P evolves according to

$$P_{k+1} = P_k + Q.$$

If $P_0 = 0$, then $P_n = nQ$, thus replicating the above uncertainty growth with n in equation (22.1).

So we can get a decent estimate for the uncertainty s_n associated with the mosquito location after n steps, namely $s_n = n\sqrt{\sigma}$. But this is incomplete information: ideally, we want to know what the *probability* is of finding a mosquito within a circle of radius R, say, surrounding the starting point after n steps. Deriving this probability is not quite 'dangerous double-bends territory'[3], but it is a bit more complex than the foregoing, so has been relegated to Appendix J.

---

[3] A nod to "The TEXbook", by Donald Knuth.

Chapter 23

CALCULUS OF VARIATIONS

WHAT, more calculus? Indeed so — although the aim here is to minimise an integral quantity. The basic idea can be most easily illustrated by means of an example:

> Suspend a heavy chain (or the empty washing line, for that matter) by its two ends. Under gravity, the chain will form a curve, but what sort of curve?

Intuitively, we might think of the arc of a circle, but in fact the answer is what is called a *hyperbolic cosine*, as will be shown below. This sort of minimisation (or maximisation, depending on context) differs from what was discussed in Section 7.7: there, the aim was to find an extremal point on a curve or a surface; now, we need to characterise an entire functional shape.

The derivation of the general equation for the extremal is quite ingenious and worth outlining here. Suppose the integral to be minimised is of the general form:

$$J = \int_a^b f\left(\frac{dy(x)}{dx}, y(x), x\right) dx, \qquad (23.1)$$

for some function f integrated between fixed limits a and b, and where y and dy/dx are functions of x, as indicated.

Imagine now, if you will, that somehow the correct extremal function $y(x)$ has been determined. Since this is the solution, any small perturbation from it will be sub-optimal. So slightly modify y everywhere except at the end-points:

$$y(x) \to y(x) + \epsilon u(x), \text{ with } u(a) = u(b) = 0,$$

where the parameter ϵ, the Greek letter 'epsilon', is a small quantity, and the function $u(x)$ is completely arbitrary — whatever it is, it will be shifting the solution away from the extremal case. In other words, the integral

$$J = \int_a^b f\left(\frac{dy(x)}{dx} + \epsilon \frac{du(x)}{dx}, \, y(x) + \epsilon u(x), \, x\right) dx,$$

will not be optimal.

At this point, on purely notational grounds, it is convenient to re-introduce the shorthand 'dash' expression for the derivative with respect to x:

$$y'(x) \text{ for } \frac{dy}{dx}, \text{ and } u'(x) \text{ for } \frac{du}{dx}.$$

The integral J may then be written in somewhat abbreviated form as

$$J = \int_a^b f\Big(y'(x) + \epsilon u'(x), \, y(x) + \epsilon u(x), \, x\Big) dx.$$

This can now be regarded as a function of ϵ and minimised in the same manner as in Section 7.7:

$$\frac{dJ}{d\epsilon} = \frac{d}{d\epsilon} \int_a^b f\Big(y'(x) + \epsilon u'(x), \, y(x) + \epsilon u(x), \, x\Big) dx,$$

$$= \int_a^b \frac{d}{d\epsilon} f\Big(y'(x) + \epsilon u'(x), \, y(x) + \epsilon u(x), \, x\Big) dx, \text{ as } a \text{ and } b \text{ are constant,}$$

$$= \int_a^b \left[u'(x)\frac{\partial f}{\partial y'} + u(x)\frac{\partial f}{\partial y}\right] dx, \text{ using equation (7.7).}$$

This does not appear to have gained very much, at least at first sight. But separate out the first term for additional consideration, revert to $u'(x) = du/dx$ and define $v(x) = \partial f/\partial y'$, so obtaining

$$\int \frac{du}{dx} v(x) dx.$$

This is now of the same form as the second term on the right-hand side of equation (7.8), except that here x is being used in place of t. Therefore,

$$\int_a^b u'(x) \frac{\partial f}{\partial y'} dx = \left[u(x)\frac{\partial f}{\partial y'}\right]_a^b - \int_a^b u(x)\frac{d}{dx}\left(\frac{\partial f}{\partial y'}\right) dx.$$

The square-bracketed quantity has to be zero, since $u(x)$ is zero at both end-points of the integral, so

$$\frac{dJ}{d\epsilon} = \int_a^b \left[-u(x)\frac{d}{dx}\left(\frac{\partial f}{\partial y'}\right) + u(x)\frac{\partial f}{\partial y}\right] dx,$$

$$= -\int_a^b u(x) \left[\frac{d}{dx}\left(\frac{\partial f}{\partial y'}\right) - \frac{\partial f}{\partial y}\right] dx.$$

To find the extremal, set $dJ/d\epsilon = 0$ as in Section 7.7. This can be achieved either by setting $b = a$ (in which case the whole problem goes away); or $u(x) = 0$ for all

23.1. SHAPE OF HEAVY CHAIN

x; or the quantity in the square brackets is zero. It was stated above that $u(x)$ is a completely arbitrary function, so a non-zero choice can be made here. This leaves the solution for the extremal curve being given by the *Euler-Lagrange* equation[1]:

$$\frac{d}{dx}\left(\frac{\partial f}{\partial y'}\right) = \frac{\partial f}{\partial y}. \tag{23.2}$$

This plays a rôle in analytical dynamics (an alternative formulation of Newton's $F = ma$, Chapter 24); finding curves of shortest distance (over the surface of a sphere for example, as used by long-distance aircraft); Einstein's General Relativity (Chapter 29); and numerous other areas.

23.1 Shape of Heavy Chain

Now for the shape of the heavy chain. The principle to be used in formulating the integral is that the chain will adopt that shape which brings its centre of gravity as near to the earth surface as possible. Suppose that the mass per unit length along the chain (or rope) is ρ (Greek 'rho'), in kilograms per metre, say, assumed constant. Thus, the total mass M of the chain will be given by

$$M = \int_0^L \rho\, ds,$$

where ds here stands for a small element of length and L is the total chain length.

Set coordinate x measured along the local earth surface and let y be the height above the ground of the element ds. To find the centre of gravity of the chain, we need the equation:

$$MgH = g\int_0^L y\rho\, ds,$$

where H is the height of the mass centre above the ground. This simply equates the total mg gravitational force on the left of the equation, to the summed gravitational force on each chain element, on the right of the equation. Therefore,

$$H = \frac{\rho}{M}\int_0^L y\, ds,$$

since ρ is a constant and g cancels out.

Now invoke Pythagoras' theorem (again) for the element ds, so that $ds = \sqrt{dx^2 + dy^2}$. This gives

$$H = \frac{\rho}{M}\int_{s=0}^L y\sqrt{dx^2 + dy^2},$$

$$= \frac{\rho}{M}\int_{x=a}^b y(x)\sqrt{1 + \left(\frac{dy}{dx}\right)^2}\, dx, \tag{23.3}$$

---

[1] Leonhard Euler, 1707 to 1783, and Joseph Louis Lagrange, 1736 to 1813.

where y is now treated as a function of x and x ranges from a to b. This is now in 'standard form', enabling the application of equation (23.2) with

$$f = y\sqrt{1 + y'^2}. \qquad (23.4)$$

That is, carrying out the two derivatives:

$$\frac{d}{dx}\left(\frac{yy'}{\sqrt{1+y'^2}}\right) = \sqrt{1+y'^2}.$$

It is not absolutely necessary to go through the detailed derivation of the form of $y(x)$, but for the interested reader the analysis is in Appendix D. We content ourselves here with the general solution:

$$y(x) = \frac{1}{C}\cosh(Cx - D), \qquad (23.5)$$

where C and D are constants, to be determined from the chain end-points. The hyperbolic cosine function cosh here was introduced in Section 7.4, but to reiterate it is formed from exponentials and is:

$$\cosh z = \frac{1}{2}\left(e^z + e^{-z}\right),$$

which has the symmetric flat-bottomed shape illustrated in Figure 23.1.

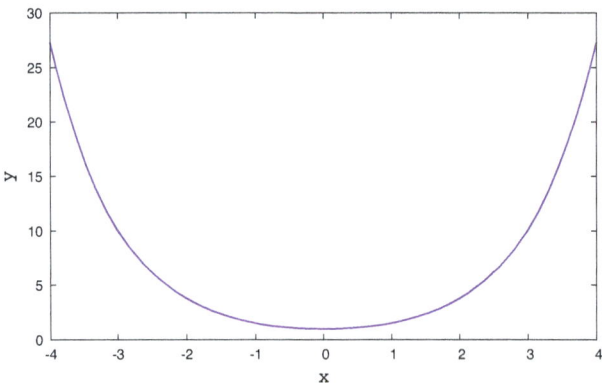

Figure 23.1: Hyperbolic cosine function

One would not have expected this particular function to form the solution.

The preceding analysis provides the generic curve shape given by equation (23.5). In any *specific* example, though, the actual length of the chain must be factored in for a complete solution. That is, there will be an additional constraint of the form

$$L = \int_{s=0}^{L} ds = \int_{x=a}^{b} \sqrt{(1 + y'^2)}\, dx,$$

23.2. GEODESICS

for a fixed value of L. This constraint can be accommodated by means of a *Lagrange multiplier*, in which the integral to be minimised is written as:

$$J = \int_{x=a}^{b} \frac{\rho}{M} y(x) \sqrt{1+y'^2}\, dx - \int_{x=a}^{b} \lambda \sqrt{(1+y'^2)}\, dx,$$

$$= \int_{x=a}^{b} \left[\frac{\rho}{M} y(x) - \lambda\right] \sqrt{(1+y'^2)}\, dx,$$

introducing the additional parameter λ. The Euler-Lagrange equation can now be applied to an augmented function f of the form

$$f = \left[\frac{\rho}{M} y(x) - \lambda\right] \sqrt{(1+y'^2)},$$

which replaces equation (23.4).

The density ρ is here specifically included within the function, to allow for a chain mass per unit length that varies with s or x; this could, for instance, be used to model one or more additional weights distributed at various points[2]. If ρ is a constant, then the chain shape is still a cosh curve but with the additional parameter λ that can be used to fix the length at L. The other two constants of integration C and D will be needed to fix the heights of the two chain end-points.

23.2 Geodesics

The term 'geodesic' is used to describe the shortest path between two points. In the familiar Euclidean flat-space geometry, this is a straight line, but in other geometries it is likely to be a curve of some form or another.

The length of the path in any geometry will be given by the integral

$$J = \int_{a}^{b} ds,$$

where ds is a small step along the path and a and b are start and end points. To find the geodesic, J needs to be minimised.

We look first at the two-dimensional flat-space situation (a sheet of paper for example), in which Pythagoras' theorem comes in again:

$$ds = \sqrt{dx^2 + dy^2}.$$

So

$$J = \int_{a}^{b} \sqrt{dx^2 + dy^2} = \int_{x_a}^{x_b} \sqrt{1+\left(\frac{dy}{dx}\right)^2}\, dx,$$

writing the integral in terms of x. Thus, referring back to equation (23.1), the function f is given by:

$$f = \sqrt{1+y'^2}.$$

---

[2] The solution inevitably becomes more complex.

On applying the Euler-Lagrange equations, (23.2), we get:

$$\frac{d}{dx}\left(\frac{y'}{\sqrt{1+y'^2}}\right) = 0, \text{ which implies that } \frac{y'}{\sqrt{1+y'^2}} = C,$$

for some constant C. Squaring both sides and rearranging results in:

$$\frac{dy}{dx} = \frac{\pm C}{\sqrt{1-C^2}} = \text{constant}.$$

That is, a straight line, as expected.

A more interesting case is to find the geodesics on the surface of a sphere of radius a [69]. This could, for instance, represent the earth[3], in which case the geodesics will give the optimal flight paths for aircraft, for example. In this case, we are dealing with three dimensions, so

$$ds = \sqrt{dx^2 + dy^2 + dz^2},$$

with the coordinate origin at the centre of the sphere. As it stands, using this ds alone to form f will again result in straight line geodesics, because no account has yet been taken of the constraint: namely, that all of the action takes place on or just above the sphere's surface. The x, y, and z coordinates are not all independent, being linked by the equation of the sphere:

$$x^2 + y^2 + z^2 = a^2.$$

This is Pythagoras' theorem yet again, and requires a revision of the integral J, so that we need to minimise the following:

$$J = \int_a^b ds, \text{ subject to the constraint } x^2 + y^2 + z^2 = a^2.$$

Accounting for the constraint *can* be achieved by expressing z in terms of x and y, but a more elegant solution works by again introducing a Lagrange multiplier λ, and reformulating J as follows:

$$J = \int_a^b \left[\sqrt{\left(\frac{dx}{d\tau}\right)^2 + \left(\frac{dy}{d\tau}\right)^2 + \left(\frac{dz}{d\tau}\right)^2} - \lambda\left(x^2 + y^2 + z^2 - a^2\right)\right] d\tau,$$

$$= \int_a^b \left[\sqrt{x'^2 + y'^2 + z'^2} - \lambda\left(x^2 + y^2 + z^2 - a^2\right)\right] d\tau,$$

here temporarily relabelling the increment of path length by τ (Greek 'tau'), to avoid confusion with s, and defining $x' = dx/d\tau$, etc.

The function f now is defined as:

$$f = \sqrt{x'^2 + y'^2 + z'^2} - \lambda\left(x^2 + y^2 + z^2 - a^2\right),$$

---

[3]Which is actually not quite spherical, being squashed in a bit at the poles, to use non-technical language.

23.2. GEODESICS

which may raise a few eyebrows due to the inclusion of a clearly zero quantity multiplying λ, but stick with it and see what happens. (The same idea is used in optimal control theory, Section 31).

Since f depends on x, y, z and their derivatives with respect to τ, there are three Euler-Lagrange equations:

$$\frac{d}{d\tau}\left(\frac{\partial f}{\partial x'}\right) = \frac{\partial f}{\partial x}, \quad \frac{d}{d\tau}\left(\frac{\partial f}{\partial y'}\right) = \frac{\partial f}{\partial y}, \quad \frac{d}{d\tau}\left(\frac{\partial f}{\partial z'}\right) = \frac{\partial f}{\partial z},$$

which give:

$$\frac{d}{d\tau}\left(\frac{x'}{\sqrt{x'^2 + y'^2 + z'^2}}\right) = -2\lambda x,$$

$$\frac{d}{d\tau}\left(\frac{y'}{\sqrt{x'^2 + y'^2 + z'^2}}\right) = -2\lambda y,$$

$$\frac{d}{d\tau}\left(\frac{z'}{\sqrt{x'^2 + y'^2 + z'^2}}\right) = -2\lambda z.$$

But in addition to these three, we must not forget the constraint, which can be obtained from $\partial f/\partial \lambda = 0$, so that:

$$x^2 + y^2 + z^2 = a^2.$$

Now revert to using $\tau = s$, so that the x', etc, are with respect to s. Then $\sqrt{x'^2 + y'^2 + z'^2} = 1$ by definition, and so

$$x'' = -2\lambda x, \quad y'' = -2\lambda y, \quad z'' = -2\lambda z, \quad \text{with } x^2 + y^2 + z^2 = a^2,$$

and where x'' stands for d^2x/ds^2, etc.

To obtain λ, differentiate the last (constraint) equation once with respect to s:

$$xx' + yy' + zz' = 0, \text{ ignoring the common factor of 2,}$$

and differentiate again:

$$(xx'' + yy'' + zz'') + \left(x'^2 + y'^2 + z'^2\right) = 0.$$

Since $\sqrt{x'^2 + y'^2 + z'^2} = 1$, we get

$$(xx'' + yy'' + zz'') + 1 = 0.$$

Now substitute $x'' = -2\lambda x$, etc, into this and rearrange to get:

$$-2\lambda\left(x^2 + y^2 + z^2\right) + 1 = 0, \text{ which implies that } 2\lambda = \frac{1}{a^2}.$$

So all of the kinematic equations for the spatial coordinates take the form:

$$x'' + \frac{x}{a^2} = 0, \quad y'' + \frac{y}{a^2} = 0, \quad z'' + \frac{z}{a^2} = 0,$$

which imply periodic solutions involving sines and cosines for each individual coordinate.

We are now in a position to determine the shape of the geodesic. Since the spherical surface looks the same from any angle, we are free to choose the coordinates in any orientation, provided the origin remains at the sphere's centre. So fix the z coordinate so that both end points a and b of the curve correspond to $z = 0$. Then solve the above differential equation for z:

$$z = A \cos\left(\frac{s}{a}\right) + B \sin\left(\frac{s}{a}\right),$$

for constants A and B. Define $s = 0$ at point a and $s = s_1$ at point b. Then, since $z = 0$ at $s = 0$, we must have $A = 0$, leaving

$$z = B \sin\left(\frac{s}{a}\right).$$

Applying the other end-point condition,

$$B \sin\left(\frac{s_1}{a}\right) = 0,$$

which means that either $B = 0$ or $s_1 = \pi a$ (more generally, $s_1 = n\pi a$ for integer n). Out of these options, the most general case will be $B = 0$, since there is no particular reason why the path length s_1 should be so constrained. But if $B = 0$, then $z = 0$ permanently, meaning that the geodesic remains in a single plane — it is a great circle path.

A more elegant approach can be found in [69], but the conclusion is the same.

With regards to geodesics on the actual earth, the WGS-84 (World Geodetic Survey 1984) defines the ellipsoidal surface as

$$x^2 + y^2 + \frac{z^2}{(1-f)^2} = R_e,$$

where R_e is the equatorial radius of the earth (6378137 metres) and f is the ellipticity (a 'flattening factor', characterising the 'squashed in at the poles' shape). In this case, the great circle path is only an approximation, and deriving the actual geodesics is quite a bit more complicated.

23.3 Maximum Area for Given Perimeter

A farmer has a certain length of fencing that he can use to enclose a flock of sheep. What is the maximum area that he can fence off for grazing, and what shape is the boundary? It is known that the answer is a circle, which gives the maximum area for a specified perimeter length, but checking this assertion by trial and error could take quite a while ...

We borrow the general expression for the area of a closed curve from Section 13.1, namely

$$A = \frac{1}{2} \oint_C (x\,dy - y\,dx)$$

23.3. MAXIMUM AREA FOR GIVEN PERIMETER

in Cartesian coordinates. Additionally, the length of available fencing is specified as L, so that

$$L = \oint_C ds.$$

As in the preceding sections, work with x as the primary coordinate and use $ds = \sqrt{dx^2 + dy^2}$. Also introduce the Lagrange multiplier λ and minimise the composite integral J:

$$J = \frac{1}{2} \oint_C (x\,dy - y\,dx) - \lambda \oint_C ds,$$
$$= \oint_C \left[\frac{1}{2}\{xy' - y\} - \lambda\sqrt{1 + y'^2} \right] dx.$$

So apply the Euler-Lagrange equation to the function f, where

$$f = \frac{1}{2}\{xy' - y\} - \lambda\sqrt{1 + y'^2}.$$

Therefore,

$$\frac{d}{dx}\left\{ \frac{1}{2}x - \frac{\lambda y'}{\sqrt{1+y'^2}} \right\} = -\frac{1}{2}.$$

This can be integrated directly, giving

$$\frac{1}{2}x - \frac{\lambda y'}{\sqrt{1+y'^2}} = -\frac{x}{2} + C,$$

for constant C. Rearrange this to obtain y':

$$y' = \frac{\pm(x - C)}{\sqrt{\lambda^2 - (x - C)^2}},$$

and integrate again to get:

$$y - D = \pm\sqrt{\lambda^2 - (x - C)^2},$$

with D another constant. Squaring both sides and rearranging then comes up with the equation for a circle in Cartesian coordinates:

$$(x - C)^2 + (y - D)^2 = \lambda^2.$$

The area encompassed will then be $\pi\lambda^2$, while the radius λ can be obtained from the known perimeter L, such that $\lambda = L/2\pi$.

In the same manner, a sphere will give the greatest volume for a specified surface area.

Chapter 24

ANALYTICAL DYNAMICS

IT may seem more than a little strange to be able to derive kinematic motions by minimising an integral, but so it is. The *principle of least action* states that the motion of mechanical systems can be obtained by minimising the integral

$$S = \int_{t_1}^{t_2} \mathcal{L}\left(q_i,\, \dot{q}_i,\, t\right) dt,$$

with t standing for time, q_i for a set of spatial coordinates and a superscript dot denoting time derivative [70].

The quantity $\mathcal{L}$ is called the *Lagrangian* of the system and can be formed from the difference of the kinetic and potential energies. In view of the calculus of variations (Section 23), the motion of the system will then be given by the Euler-Lagrange equations:

$$\frac{d}{dt}\left(\frac{\partial \mathcal{L}}{\partial \dot{q}_i}\right) = \frac{\partial \mathcal{L}}{\partial q_i}, \tag{24.1}$$

for each of the i spatial components.

Let us try the above scheme for the case of a planetary orbit, assuming that the motion is in a plane so that only two spatial components are necessary. The kinetic and potential energies are then as follows:

$$\text{Kinetic energy: } \frac{1}{2}m\left\{\dot{x}^2 + \dot{y}^2\right\},$$

$$\text{Potential energy: } -\frac{GMm}{r},$$

where $\dot{x}$ stands for dx/dt, etc, GM is the product of the gravitational constant and the mass of the sun, and m is the mass of the planet. The quantity $r = \sqrt{x^2 + y^2}$, with both x and y measured from the centre of the sun.

It can be seen that the kinetic energy has a positive sign, whereas the gravitational potential energy is *negative*. This leads to the possibility that the total energy of a system could be zero, and there is some acceptance among cosmologists that this may well apply to the universe as a whole [71].

In the present case, the Lagrangian is given by:

$$\mathcal{L} = \text{kinetic energy} - \text{potential energy} = \frac{1}{2} m \{\dot{x}^2 + \dot{y}^2\} + \frac{GMm}{r},$$

and the two Euler-Lagrange equations are:

$$\frac{d}{dt}\left(m \frac{dx}{dt}\right) = \frac{\partial}{\partial x}\left(\frac{GMm}{r}\right),$$

$$\frac{d}{dt}\left(m \frac{dy}{dt}\right) = \frac{\partial}{\partial y}\left(\frac{GMm}{r}\right).$$

If m is a constant, it drops out of the equations and we end up with

$$\frac{d^2 x}{dt^2} = -\frac{GM}{r^2} x,$$

$$\frac{d^2 y}{dt^2} = -\frac{GM}{r^2} y,$$

using $r^2 = x^2 + y^2$ to obtain $\partial/\partial x(1/r)$, etc. These are now the same as in Section 11.4, although here derived using the minimisation of an integral rather than formed from Force = mass × acceleration.

Return now to equation (24.1) and define the generalised momentum p_i:

$$p_i = \frac{\partial \mathcal{L}}{\partial \dot{q}_i}, \qquad (24.2)$$

so that equation (24.1) may be written as:

$$\dot{p}_i = \frac{\partial \mathcal{L}}{\partial q_i}. \qquad (24.3)$$

Also define the *Hamiltonian*[1]:

$$H = \sum_j \dot{q}_j \frac{\partial \mathcal{L}}{\partial \dot{q}_j} - \mathcal{L}, \qquad (24.4)$$

$$= \sum_j p_j \dot{q}_j - \mathcal{L}. \qquad (24.5)$$

Partially differentiating this with respect to q_i then gives:

$$\frac{\partial H}{\partial q_i} = \frac{\partial}{\partial q_i}\left[\sum_j p_j \dot{q}_j - \mathcal{L}\right] = -\frac{\partial \mathcal{L}}{\partial q_i},$$

---

[1] After William Rowan Hamilton, 1805 to 1865.

since the summation term is independent of $\dot{q}_i$ explicitly.

In view of equation (24.3), we then get:

$$\dot{p}_i = -\frac{\partial H}{\partial q_i}.$$

Now partially differentiate H with respect to p_i:

$$\frac{\partial H}{\partial p_i} = \frac{\partial}{\partial p_i}\left[\sum_j p_j \dot{q}_j - \mathcal{L}\right] = \dot{q}_i,$$

since $\mathcal{L}$ is regarded as a function of $q_i, \dot{q}_i$ and t (and independent of p_i).

We thus end up with Hamilton's equations:

$$\dot{q}_i = \frac{\partial H}{\partial p_i},$$

$$\dot{p}_i = -\frac{\partial H}{\partial q_i},$$

which form an equivalent set of kinematic equations to the Euler-Lagrange ones.

The above excursion into H may seem like mathematical manipulation for the sake of it, but the quantity H reappears in optimal control theory (Chapter 31) and is also relevant in quantum mechanics (there treated as an operator; see Section 30).

Suppose we now take the ordinary derivative of H from equation (24.5) with respect to time:

$$\frac{dH}{dt} = \sum_j p_j \ddot{q}_j + \sum_j \dot{p}_j \dot{q}_j - \frac{d\mathcal{L}}{dt}.$$

But

$$\frac{d\mathcal{L}}{dt} = \frac{d}{dt}\mathcal{L}(q_i, \dot{q}_i, t) = \sum_j \frac{\partial \mathcal{L}}{\partial q_j}\dot{q}_j + \sum_j \frac{\partial \mathcal{L}}{\partial \dot{q}_j}\ddot{q}_j + \frac{\partial \mathcal{L}}{\partial t},$$

$$= \sum_j \dot{p}_j \dot{q}_j + \sum_j p_j \ddot{q}_j + \frac{\partial \mathcal{L}}{\partial t}, \text{ using equations (24.3) and (24.2).}$$

Putting the pieces together, we find that:

$$\frac{dH}{dt} = -\frac{\partial \mathcal{L}}{\partial t}.$$

What this means is that if $\mathcal{L}$ does not contain time explicitly (*i.e.* is a function only of q_i and $\dot{q}_i$), then H must be a constant of the motion — which can be very useful when solving the equations. The quantity H is interpreted as the energy of the system.

A closely related result here is *Noether's theorem*[2]: she was the first to show that physical symmetries relate to conserved quantities. For example, if the Lagrangian is unchanged by a time shift or time reversal, energy is conserved. Similarly, spatial homogeneity corresponds to conservation of momentum, while spatial isotropy corresponds to conservation of angular momentum.

---

[2] Amelie Emmy Noether, 1882 to 1935.

Chapter 25

ASSORTED ENGINEERING APPLICATIONS

MATHEMATICS plays very big part across all the various domains of engineering, and the examples here are only a tiny subset.

25.1 Hydrostatics

This section takes a look at the mathematics of fluid statics, and in particular how the fluid pressure increases with depth. To take a somewhat homely example, what is the force on the sides of a large rectangular fish tank? Or, scaling the problem up a bit, what is the force on the retaining wall of a hydroelectric dam?

To assess the situation, assume that the fluid is incompressible (near enough valid for water) and look at a small (imaginary) rectangular box somewhere inside the overall volume; this box is illustrated in Figure 25.1.

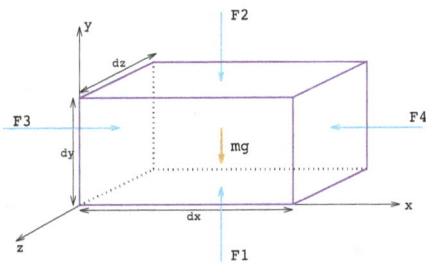

Figure 25.1: Balance of forces on a small fluid 'box'

The sides of the box are designated dx, dy and dz in the three coordinate directions, and we are interested in the balance of forces on this small box of fluid alone.

The only external force acting on this box is gravity, denoted mg in Figure 25.1 (see Section 11.1 for the force due to gravity). All of the other forces involved stem from the presence of the rest of the fluid outside the box, which may be considered as exerting a pressure on each of the six sides. The box itself is not moving, so all of these forces, internal and external, must balance out.

So on the top face, there is a force F_2 acting downwards, and on the bottom face a force F_1 acting upwards. Similarly, sideways forces F_3 and F_4 act laterally; the forces in the z-direction (out of the page) are suppressed to clarify the diagram.

The balance of forces in the x and y directions then must be:

$$F_3 = F_4,$$
$$F_1 = F_2 + mg.$$

But what is the force on one of the box faces? The quantity of interest here is pressure p, defined as force per unit area, so force will equal to pressure times area. Therefore, on the bottom face:

$$F_1 = p\,dx\,dz,$$

and on the left face:

$$F_3 = p\,dy\,dz.$$

The box is assumed small enough that the pressure at the bottom left corner is essentially the same as on the bottom and left faces. Now look at the pressure on the right face. This will be, approximately,

$$p + \frac{\partial p}{\partial x}dx,$$

the partial derivatives occurring because pressure is assumed, at this stage of the analysis, to be a function of x, y and z. Therefore,

$$F_4 = \left(p + \frac{\partial p}{\partial x}dx\right)dy\,dz.$$

The balance of forces in the x-direction then gives:

$$p\,dy\,dz = \left(p + \frac{\partial p}{\partial x}dx\right)dy\,dz, \text{ which implies that } \frac{\partial p}{\partial x} = 0.$$

In a very similar manner, it is found that $\partial p/\partial z = 0$, so it can be inferred that p is actually only a function of y — it changes only with depth (as might have been anticipated).

25.1. HYDROSTATICS

It remains to look at the balance of forces in the y direction, which — from above — will give:

$$p\,dx\,dz = \left(p + \frac{dp}{dy}dy\right)dx\,dz + mg.$$

The final step is to replace the mass m of fluid in the box with density ρ (defined as mass per unit volume) times volume. That is,

$$p\,dx\,dz = \left(p + \frac{dp}{dy}dy\right)dx\,dz + \rho g\,dx\,dy\,dz.$$

Cancelling the common terms and rearranging:

$$\frac{dp}{dy} = -\rho g, \qquad (25.1)$$

which is a key equation in hydrostatics. The above derivation is deliberately long-winded to show how simple 'common-sense' rules plus the application of calculus basics can lead to something practical.

If (as in many cases of practical interest) the fluid density is a constant (as in water all the way down), then ρg may be treated as a constant, so that the pressure difference between two points will be given by the simple relation:

$$p_2 - p_1 = -\rho g\,(y_2 - y_1). \qquad (25.2)$$

If, then, $p_1 = p_0$ the pressure (typically atmospheric) at the fluid surface, and h is measured downwards from that surface, then the pressure at depth h will be given by:

$$p = p_0 + \rho g h. \qquad (25.3)$$

This can be applied directly to determine the pressure of water at any point on the tank sidewall.

To find the *total* vector force $\mathbf{F}$ on a submerged object, it is necessary to integrate the pressure over its surface area. The following equation is then applicable:

$$\mathbf{F} = \int_A p\,d\mathbf{A} = \int_A (p_0 + \rho g h)\,d\mathbf{A}, \qquad (25.4)$$

where A is the area of interest and $d\mathbf{A}$ is the differential area (itself a vector).

It is now possible to apply equation (25.4) to the force of the water on the vertical side of a fish tank, of depth H and width W (in the z direction). In this case, the atmospheric pressure p_0 can be ignored, since it is the same both outside and inside the tank. So,

$$F_x = \int_A \rho g h\,dA = \rho g W \int_{y=0}^{H} y\,dy = \frac{1}{2}\rho g W H^2, \qquad (25.5)$$

and this force is in the x direction. It may be seen that this involves the *square* of the depth, and so increases markedly as the tank is made deeper. To put some sample numbers in, suppose the tank is 1 metre deep and wide; then, with $\rho = 1000$ kilograms per cubic metre and $g = 9.8$ metres per second squared, we get $F_x = 4900$ Newtons (the unit of force).

The same equation (25.5) applies to a vertical dam wall.

Things get a bit more interesting if the fluid is in motion. For instance, suppose that the fish tank has a small hole on one side, near the bottom, that allows the water to escape out into atmospheric pressure. In this case, the key equation is from Bernoulli[1], and brings in the speed of fluid flow, v:

$$\frac{1}{2}\rho v^2 + p + \rho y g = C \text{ (a constant)}.$$

This is essentially a consequence of Newton's law: force = mass × acceleration. The density ρ and acceleration due to gravity g are assumed constant. It can be seen that for *static* flow, with $v = 0$, this reduces to equation (25.2).

The tank with a hole at the base is illustrated in Figure 25.2, with the symbol 'S' standing for surface conditions, and 'E' for exit conditions with outflow speed v.

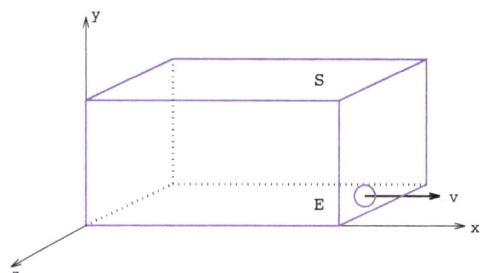

Figure 25.2: Fish tank with a hole at the base

Assume atmospheric pressure p_0 at S, and that the outflow of water does not noticeably disturb the surface, so $v = 0$ here. Therefore at S, with $y = y_s$, we have:

$$p_0 + \rho y_s g = C, \text{ thus giving } C.$$

At E with $y = y_e$, since this gap is now open to the atmosphere, $p = p_0$ again, so

$$\frac{1}{2}\rho v^2 + p_0 + \rho y_e g = C,$$
$$= p_0 + \rho y_s g.$$

---
[1] Daniel Bernoulli, 1700 to 1782.

25.2. VIBRATING STRINGS

Solving this then gives an equation for the speed of water outflow, v:

$$v^2 = 2g(y_s - y_e). \tag{25.6}$$

The deeper the hole is relative to the surface, the faster the outflow, which might have been expected. It can also be inferred that the speed drops with time, as the level of water falls in the tank.

Continuing with the same example, now ask how long it will take for the tank to empty? (or, more specifically, for the water level to fall from $y = y_s$ to $y = y_e$). To answer this question, we need to look at rates of change of volumes of fluid. Start by writing equation (25.6) in the form:

$$\frac{dx}{dt} = \sqrt{2g(y - y_e)}, \tag{25.7}$$

since v in Figure 25.2 is positive in the x-direction, and noting that y is variable.

Next suppose that the outflow hole connects seamlessly with a horizontal pipe of cross-sectional area a. A small element of fluid volume progressing along this pipe will then have volume given by $a\,dx$.

Since the fluid is incompressible, this small volume element must be the same as has just been extracted from the tank. So let the tank cross-sectional area be A (looking down on it). The volume of fluid lost due to $a\,dx$ must then be equivalent to $A\,dy$, dy being the corresponding thickness of a horizontal slice of fluid in the tank. Thus,

$$A\,dy = a\,dx.$$

Divide both sides by a small time slice dt:

$$A\frac{dy}{dt} = -a\frac{dx}{dt},$$
$$= -a\sqrt{2g(y - y_e)}, \text{ from equation (25.7).}$$

The minus sign is introduced because a positive dx/dt corresponds to a *decrease* in y, as the water level drops in the tank.

This equation now links time to the rate of drop of water level in the tank, so the total time T for the water level to fall from y_s to y_e is given by:

$$T = -\frac{A}{a\sqrt{2g}} \int_{y=y_s}^{y_e} \frac{dy}{\sqrt{y - y_e}},$$
$$= \frac{2A\sqrt{y_s - y_e}}{a\sqrt{2g}}.$$

From this, it may be deduced that if the outflow hole is small (meaning small a), so the time T becomes correspondingly large, which is what we would expect.

25.2 Vibrating Strings

One of the simplest applications of partial differential equations is in the description of the motion of a plucked string, as in a guitar or violin, for instance.

Suppose that the string is under tension (it won't give any useful tunes otherwise), and not plucked too hard, so that the deviations from linearity are small. Then introduce the axes shown in Figure 25.3., with the string stretched from $x = 0$ to $x = L$ and write the lateral displacement as u. Left to itself, u would be zero but in Figure 25.3 the string has been pulled up a bit.

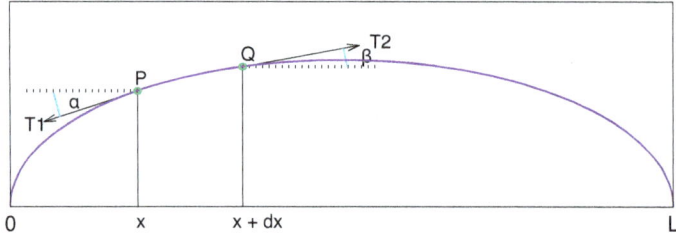

Figure 25.3: Vibrating string

Then look at two closely-spaced points P and Q, with P at x and Q at $x + dx$, for small dx. The string is under tension, so these points are being pulled apart, with tangential tension forces T_1 acting to the left and T_2 acting to the right. These forces are assumed to balance in the x direction, so the string only moves up and down. Therefore,

$$T_1 \cos \alpha = T_2 \cos \beta = T,$$

T being a constant and where the angles α, β are with respect to the horizontal.

In the vertical u direction, however, the string is accelerating. Using Newton's Force = mass × acceleration equation, the balance of forces in u will be given by:

$$T_2 \sin \beta - T_1 \sin \alpha = \rho\, dx \frac{d^2 u}{dt^2},$$

where ρ is the mass per unit length of the string, the small string length in question is dx and d^2u/dt^2 is the acceleration.

Combining both equations, we get:

$$\frac{T_2 \sin \beta}{T_2 \cos \beta} - \frac{T_1 \sin \alpha}{T_1 \cos \alpha} = \frac{\rho\, dx}{T} \frac{d^2 u}{dt^2},$$

which is the same as:

$$\tan \beta - \tan \alpha = \frac{\rho\, dx}{T} \frac{d^2 u}{dt^2}.$$

But $\tan \alpha$ is just the slope of the tangent to the string at P, and similarly for Q. So, equivalently,

$$\left.\frac{\partial u}{\partial x}\right|_{x+dx} - \left.\frac{\partial u}{\partial x}\right|_{x} = \frac{\rho\, dx}{T} \frac{\partial^2 u}{\partial t^2},$$

25.2. VIBRATING STRINGS

here switching over to partial derivatives, since u is a function of both x and t. Now divide both sides by dx; then in the limit as $dx \to 0$, the left-hand-side of the equation will tend to the second derivative of u with respect to x. That is,

$$\frac{\partial^2 u}{\partial t^2} = c^2 \frac{\partial^2 u}{\partial x^2}, \text{ where } c^2 = \frac{T}{\rho}. \tag{25.8}$$

This is the wave equation in one spatial dimension x, and the quantity c has units of speed.

The same equation describes the propagation of electromagnetic waves in a vacuum, in one dimension, c then being the speed of light.

Now to provide a solution to equation (25.8) for the case of the string under tension. Clearly, u has to be zero at $x = 0$ and $x = L$, since it is pinned at these points and these suffice for setting the boundary conditions. For an initial condition at $t = 0$, we suppose that the string is pulled up into a shape given by $u = g(x)$, leaving the function g undefined for the time being. Also at $t = 0$, assume that the rate of change of u with respect to t is zero — that is, at $t = 0$, the string has been pulled into some shape and held there, instantaneously static.

We try what is known as a *separable solution*, in that

$$u(x,t) = X(x)T(t),$$

namely a product of two functions, each of which is a function of only one variable. This is often worth trying as a first stab, but whether the approach works in a particular situation depends very much on the initial and boundary conditions; an example of where separability is not possible can be found in Section 25.3 below.

So put this form of u into equation (25.8) and get the following:

$$X(x)\frac{d^2 T}{dt^2} = c^2 T(t)\frac{d^2 X}{dx^2}.$$

Divide both sides by $X(x)T(t)$:

$$\frac{1}{T(t)}\frac{d^2 T}{dt^2} = \frac{c^2}{X(x)}\frac{d^2 X}{dx^2}.$$

This says that there is a function of t alone on the left-hand-side, which is equal to a function of x alone on the right-hand-side. Both expressions must therefore be equal to a constant k, say; this preserves the independence of the underlying variables x and t. So we write

$$\frac{1}{T(t)}\frac{d^2 T}{dt^2} = \frac{c^2}{X(x)}\frac{d^2 X}{dx^2} = k,$$

which gives the following pair of ordinary differential equations:

$$\frac{d^2 T}{dt^2} = kT(t), \tag{25.9}$$

$$\frac{d^2 X}{dx^2} = \frac{k}{c^2}X(x). \tag{25.10}$$

Look first at the equation for $X(x)$. As it stands, this will give solutions of the form:
$$X(x) = Ae^{x\sqrt{k}/c} + Be^{-x\sqrt{k}/c},$$

for constant A and B. However, applying the boundary conditions at $x = 0$ and $x = L$ then results in $X(x) = 0$ for all x, which is not very helpful. So define k to be *negative*, such that $k = -\nu^2$, in which case
$$X(x) = A\cos\left(\frac{\nu x}{c}\right) + B\sin\left(\frac{\nu x}{c}\right),$$

again with constants A and B. This now more promising, since the boundary conditions give $A = 0$ (which is acceptable), together with
$$B\sin\left(\frac{\nu L}{c}\right) = 0.$$

We can either choose $B = 0$, which ends up with the trivial (and unacceptable) solution $X = 0$ as before; or else fix ν such that
$$\frac{\nu L}{c} = n\pi, \text{ for integer } n > 0.$$

Negative values of n are mirror images of the positive values and need not be included. Also, $n = 0$ gives the trivial solution again. But apart from these remarks, no further information is available concerning the specific value of n, so the best we can do is to write
$$X(x) = B_n \sin\left(\frac{n\pi x}{L}\right),$$

for arbitrary integer n and linking the constant B_n to its companion frequency.
Having obtained some information on k, return now to equation (25.9), which becomes:
$$\frac{d^2 T}{dt^2} = -\nu^2 T, \quad \nu = \frac{n\pi c}{L}.$$

Thus T also exhibits sinusoidal-type solutions in time:
$$T(t) = C_n \cos\nu t + D_n \sin\nu t.$$

So a single candidate solution to u must have the form
$$u_n(x,t) = \sin\left(\frac{n\pi x}{L}\right)\left[C'_n \cos\nu t + D'_n \sin\nu t\right],$$

here absorbing B_n into C_n and D_n to create C'_n, D'_n (which are only constants after all).

For a *complete* solution, a sum over all values of n is appropriate, so that
$$u(x,t) = \sum_{n=1}^{\infty} \sin\left(\frac{n\pi x}{L}\right)\left[C'_n \cos\nu t + D'_n \sin\nu t\right],$$

25.2. VIBRATING STRINGS

and it is now possible to factor in the initial condition that $u(x,0) = g(x)$, which results in:

$$\sum_{n=1}^{\infty} C'_n \sin\left(\frac{n\pi x}{L}\right) = g(x).$$

With reference to Chapter 16, this is just the Fourier sine representation of the function $g(x)$, and can be used to define unique values for C'_n. We have left $g(x)$ undefined (apart from being zero at both ends of the string), so this is as far as we wish to proceed.

The other initial condition was that $\partial u/\partial t = 0$ at $t = 0$, and on applying this we find that $D'_n = 0$. Therefore, the complete solution is given by:

$$u(x,t) = \sum_{n=1}^{\infty} C'_n \sin\left(\frac{n\pi x}{L}\right) \cos \nu t, \quad \nu = \frac{n\pi c}{L}.$$

This is not quite the end of the analysis, though, since some interesting physics can be extracted from this equation. To do so, we use the relation:

$$2 \sin\left(\frac{S+T}{2}\right) \cos\left(\frac{S-T}{2}\right) = \sin S + \sin T, \quad \text{for any } S \text{ and } T.$$

(similar equations exist for products of sines and products of cosines). Therefore,

$$u(x,t) = \sum_{n=1}^{\infty} C'_n \left[\sin\left\{\frac{n\pi}{L}(x+ct)\right\} + \sin\left\{\frac{n\pi}{L}(x-ct)\right\}\right]. \quad (25.11)$$

This now written as a sum of an *advanced wave* involving $x+ct$ and a *retarded wave* involving $x-ct$. Since equation (25.8) also occurs in electromagnetic theory, it can be expected that similar solutions arise there also — both are entirely valid mathematically. However, the advanced wave — which propagates into the past — is generally dropped as being unphysical, leaving the retarded wave (propagating into the future) as the only valid solution. An interesting exception to the assumption that the future cannot affect the present[2] may be found in [72], which provides a transactional interpretation of quantum mechanics and proposes an alternative understanding of quantum entanglement.

As a final remark here, equation (25.8) admits the following solution:

$$u(x,t) = f(x-ct) + g(x+ct), \quad (25.12)$$

where f and g are any twice-differentiable functions (although subject to satisfying the initial and boundary conditions for a complete solution). Equation (25.11) is of this form.

---

[2] "The future ain't like it used to be." (Yogi Berra)

25.3 Heat Transfer

The equation for the flow of heat in a solid or a static fluid is given by the diffusion equation:

$$\frac{\partial T}{\partial t} = c^2 \frac{\partial^2 T}{\partial x^2}, \tag{25.13}$$

here in terms of time t and one spatial dimension x. T is the temperature, and the parameter c^2 is defined by:

$$c^2 = \frac{K}{\sigma \rho},$$

where K is the thermal conductivity, ρ the density and σ the specific heat. We now apply this equation to the situation illustrated in Figure 25.4.

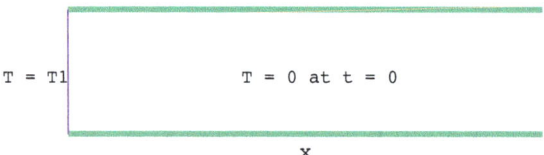

Figure 25.4: Heat flow in long insulated bar

Initially, at $t = 0$, the temperature throughout is zero. The upper and lower green boundaries are assumed to be perfect insulators, so the heat flow is in x alone; the bar is deemed to extend indefinitely off to the right. At $x = 0$, a heat source of temperature T_1 is applied (and assuming $T_1 > 0$). How then does the temperature in the bar vary with time and distance?

As a brief aside, for a more general application, the insulating boundaries can be modelled mathematically by $\partial T/\partial y = 0$, where y is in the vertical direction. In the present case, there is assumed to be no variation of temperature with y.

Now apply a Laplace transform (from Chapter 20) in the time domain to equation (25.13), defining:

$$\Phi(x) = \int_{t=0}^{\infty} e^{-st} T(x,t) dt.$$

Therefore,

$$\int_{t=0}^{\infty} e^{-st} \frac{\partial T}{\partial t} dt = c^2 \int_{t=0}^{\infty} e^{-st} \frac{\partial^2 T}{\partial x^2} dt,$$

$$= c^2 \frac{\partial^2}{\partial x^2} \int_{t=0}^{\infty} e^{-st} T(x,t) dt,$$

$$= c^2 \frac{d^2 \Phi(x)}{dx^2}, \text{ since } \Phi \text{ is a function of } x \text{ alone.}$$

25.3. HEAT TRANSFER

Evaluating the integral on the left-hand side of this equation by parts (Section 7.10) gives

$$\int_{t=0}^{\infty} e^{-st}\frac{\partial T}{\partial t}dt = \left[e^{-st}T(x,t)\right]_0^{\infty} + s\int_0^{\infty} e^{-st}T(x,t)dt,$$

$$= s\Phi(x), \text{ since } T = 0 \text{ at } t = 0.$$

The differential equation for Φ now becomes:

$$\frac{d^2\Phi(x)}{dx^2} = \frac{s}{c^2}\Phi(x),$$

which will have solutions of the form

$$\Phi(x) = Ae^{x\sqrt{s}/c} + Be^{-x\sqrt{s}/c},$$

for constants A and B. The increasing exponential term can be dropped, since this implies a temperature growing with x (contrary to experience and expectation). To determine B, the boundary condition at $x = 0$ needs to be applied to Φ, resulting in:

$$\Phi(0) = \int_{t=0}^{\infty} e^{-st}T(0,t)dt = \int_{t=0}^{\infty} e^{-st}T_1 dt = \frac{T_1}{s}.$$

So the complete solution for $\Phi(x)$ is:

$$\Phi(x) = \frac{T_1}{s}e^{-x\sqrt{s}/c}.$$

We now need to find the inverse of this, to map back to the time domain, and recourse to a table of Laplace transforms results in:

$$T(x,t) = T_1 \operatorname{erfc}\left(\frac{x}{2c\sqrt{t}}\right), \qquad (25.14)$$

where the *complementary error function* erfc(z) is defined for argument z as:

$$\operatorname{erfc}(z) = 1 - \frac{2}{\sqrt{\pi}}\int_0^z e^{-\xi^2}d\xi, \text{ with } \operatorname{erfc}(z) \to 0 \text{ as } z \to \infty \text{ and } \operatorname{erfc}(0) = 1.$$

It can be seen that this solution cannot be separated out into a function of x multiplied by a function of t, so the approach used in Section 25.2 above is inapplicable here. Indeed, if we try $T(x,t) = X(x)H(t)$, say, and separate the variables, it is found that $H(t) = H_0 e^{-kt}$ (with $k > 0$ on physical grounds). Applying the initial condition then requires that $H_0 = 0$.

The easiest way to see the spread of temperature with x and t is graphically, and the profiles of T with respect to x for a discrete set of times is shown in Figure 25.5. Here, $T_1 = 100$ degrees and c has been set (completely arbitrarily) to unity.

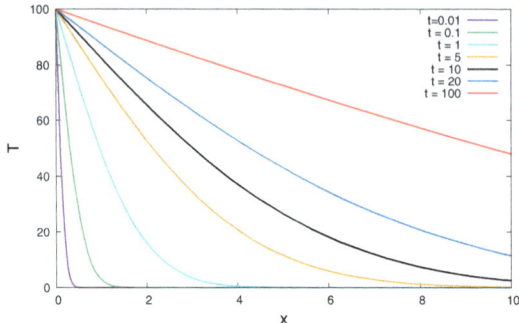

Figure 25.5: Temperatures in long insulated bar

It can be seen that even for very small times, there is diffusion of heat some way along the bar; in fact, even out to large values of x, albeit at infinitesimal levels. At longer times, the heat rises more and more along the length of the bar, which is what would be expected.

25.4 Vibration of a Beam

An equation for the vibration of a uniform solid beam can be found in [6]:

$$\frac{\partial^2 u}{\partial t^2} + c^2 \frac{\partial^4 u}{\partial x^4} = 0, \quad c^2 = \frac{EI}{\rho A}, \tag{25.15}$$

where E is Young's modulus of elasticity, I is the moment of inertia, ρ is density and A is the cross-sectional area. Since my policy has (mostly) been to explain where equations come from, a brief explanation may be found at the end of this section. The fourth-order derivative really does call for something more than a bit of hand-waving.

The beam itself is illustrated in Figure 25.6, with vibrations assumed to occur in the vertical u direction only. We can imagine this structure being firmly clamped at the left-hand end, with the right-hand end free to move up and down. Such a structure is known as a *cantilever*; some stair treads are made to this design, and the architect would prefer to know what happens when a child jumps up and down on the end.

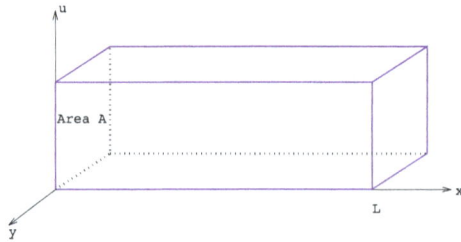

Figure 25.6: Vibration of a beam

25.4. VIBRATION OF A BEAM

The cross-sectional area refers to the area in the y-u plane, while the moment of inertia can be regarded as characterising mass movement around the y axis as a pivot.

The vertical displacement $u(x,t)$ is a function of length x along the beam and time t. We assume that, initially, the beam is bent upwards or downwards into some shape, so that:

$$u(x,0) = f(x), \text{ (the shape left indeterminate for the time being)},$$

and held static at that time, so that

$$\left.\frac{\partial u(x,t)}{\partial t}\right|_{t=0} = 0.$$

At the left-hand end, the beam cannot move and cannot twist (think 'embedded into the wall'), so:

$$u(0,t) = 0 \text{ and } \left.\frac{\partial u(x,t)}{\partial x}\right|_{x=0} = 0. \tag{25.16}$$

At the other unsupported end of the beam, at $x = L$, we impose the conditions that its curvature and rate of change of curvature must be zero:

$$\left.\frac{\partial^2 u(x,t)}{\partial x^2}\right|_{x=L} = 0 \text{ and } \left.\frac{\partial^3 u(x,t)}{\partial x^3}\right|_{x=L} = 0. \tag{25.17}$$

(Recall from Section 7.1 that the slope is given by the first derivative (which in this example can change at $x = L$); the second derivative then gives the rate of change of slope, namely curvature).

For the solution, assume separability so that:

$$u(x,t) = X(x)T(t),$$

which implies that

$$\frac{1}{T}\frac{d^2T}{dt^2} = -\frac{c^2}{X}\frac{d^4X}{dx^4} = -k^2,$$

for some constant k. The choice of sign on k^2 reflects the expectation that T will be oscillatory in time. Therefore,

$$\frac{d^2T}{dt^2} + k^2 T = 0,$$
$$\frac{d^4X}{dx^4} - \frac{k^2}{c^2}X = 0. \tag{25.18}$$

The solution for T is, simply,

$$T(t) = \alpha \cos kt + \beta \sin kt,$$

for constant α and β. The vibration frequency k is at this stage completely undefined, and we will need to solve for X to go any further. Thus try a generic form for X in the form of $e^{\lambda x}$ and substitute into equation (25.18), so that

$$\lambda^4 = \frac{k^2}{c^2}.$$

Taking the square root of this:

$$\lambda^2 = \pm\frac{k}{c}.$$

Treating the positive and negative roots individually, we get:

Positive root: $\lambda = +\sqrt{\frac{k}{c}}$ or $-\sqrt{\frac{k}{c}}$,

Negative root: $\lambda = +i\sqrt{\frac{k}{c}}$ or $-i\sqrt{\frac{k}{c}}$,

using $i^2 = -1$ (see Chapter 17).
So $X(x)$ takes the general form:

$$X(x) = A\cos\nu x + B\sin\nu x + C\cosh\nu x + D\sinh\nu x,$$

where, to save a bit on the typing, ν is defined as:

$$\nu = \sqrt{\frac{k}{c}},$$

and the hyperbolic cosh and sinh functions are defined in Section 23.1.

It is now possible to take account of the boundary conditions at $x = 0$ and $x = L$, for which the following derivatives of X will be needed:

$$\frac{dX}{dx} = \nu\left[-A\sin\nu x + B\cos\nu x + C\sinh\nu x + D\cosh\nu x\right],$$

$$\frac{d^2X}{dx^2} = \nu^2\left[-A\cos\nu x - B\sin\nu x + C\cosh\nu x + D\sinh\nu x\right],$$

$$\frac{d^3X}{dx^3} = \nu^3\left[A\sin\nu x - B\cos\nu x + C\sinh\nu x + D\cosh\nu x\right],$$

where the derivatives of the hyperbolic functions in Section 23.1 have been employed.

Imposing the boundary conditions at $x = 0$ from equation (25.16) then requires that:

$$A + C = 0,$$
$$\nu(B + D) = 0,$$

which together imply that $C = -A$ and $D = -B$; an alternative option for the second equation — that $\nu = 0$ — is rejected since $k = 0$ does not result in the expected sinusoidal behaviour of T.

25.4. VIBRATION OF A BEAM

At the right hand boundary at $x = L$, we must also have

$$-A\cos\nu L - B\sin\nu L - A\cosh\nu L - B\sinh\nu L = 0,$$
$$A\sin\nu L - B\cos\nu B - A\sinh\nu L - B\cosh\nu L = 0,$$

having substituted for C and D. These then provide two simultaneous equations for A and B:

$$A(\cos\nu L + \cosh\nu L) + B(\sin\nu L + \sinh\nu L) = 0,$$
$$A(\sin\nu L - \sinh\nu L) - B(\cos\nu L + \cosh\nu L) = 0.$$

Use the second of these to get B in terms of A:

$$B = \frac{(\sin\nu L - \sinh\nu L)}{(\cos\nu L + \cosh\nu L)} A,$$

and substitute into the first equation:

$$A(\cos\nu L + \cosh\nu L) + (\sin\nu L + \sinh\nu L)\frac{(\sin\nu L - \sinh\nu L)}{(\cos\nu L + \cosh\nu L)} A = 0.$$

At first sight, this might lead us to believe that $A = 0$; this is certainly a solution to this algebraic equation, but hardly beneficial, since then $B = 0$ and $X = 0$ as well. A rather more useful solution is given by

$$(\cos\nu L + \cosh\nu L)^2 + (\sin\nu L + \sinh\nu L)(\sin\nu L - \sinh\nu L) = 0,$$

which leaves A indeterminate. Multiplying out the terms and using the known identities $\cos^2 x + \sin^2 x = 1$, $\cosh^2 x - \sinh^2 x = 1$, we finally get (see [6]):

$$\cos\nu L \cosh\nu L = -1. \qquad (25.19)$$

This provides values for ν in terms of L and thereby the expected oscillation frequencies of the bar. As to what the frequencies actually are, it helps to define $\xi = \nu L$ and then plot the function

$$g(\xi) = 1 + \cos\xi \cosh\xi,$$

with ξ along the axis, as in Figure 25.7.

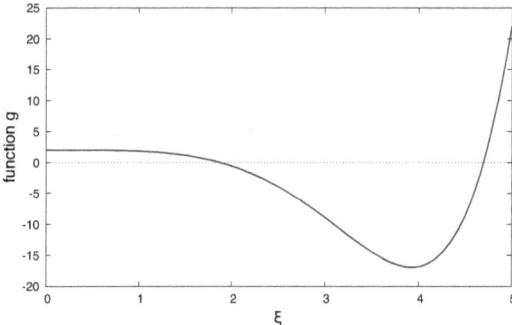

Figure 25.7: Initial values of $g(\xi)$

This is not actually the best way of graphically estimating the roots, since as ξ increases the value of $\cosh \xi$ grows massively and g swings so much between large positive and negative values that any detail is rapidly lost. So we exploit the fact that $\cosh \xi$ is always positive and define a new function:

$$h(\xi) = \cos \xi + \frac{1}{\cosh \xi}.$$

This will have exactly the same set of zeros as $g(\xi)$, but plotting it gives the much more informative picture in Figure 25.8.

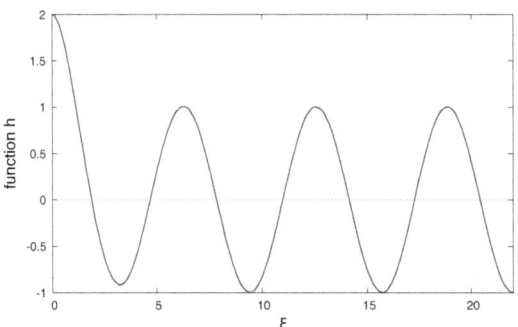

Figure 25.8: Values of $h(\xi)$

Apart from the region near to $\xi = 0$, the function becomes close to a sinusoid in behaviour, with a very regular set of evenly-spaced roots. Looking back at the definition of g, we can appreciate that as ξ increases, $\cosh \xi$ will be large and must be balanced by $\cos \xi \approx 0$ in order to achieve $g = 0$. So we expect that as ξ increases, the zeros of g or h will be given more and more accurately by the simple relation

$$\xi = \frac{n\pi}{2}, \text{ for odd values of } n \text{ (namely } n = 1, 3, 5, 7, \ldots \text{) and } n > \text{about } 3.$$

In fact, the lowest frequency has a value given by

$$\xi = \nu L \approx 1.875,$$

for which the corresponding k is:

$$k \approx \frac{3.516c}{L^2}.$$

Which means that long beams have lower oscillation frequencies than shorter ones. Such information might be useful in ensuring that people don't feel seasick when going up and down cantilevered stairs...

It is also quite obvious from Figure 25.8 that there are infinitely many possible oscillation frequencies.

A very similar separable-solutions approach to the above can be adopted if the beam is clamped at both ends, which is the more usual configuration for a stair

25.4. VIBRATION OF A BEAM

tread. In this case, the boundary conditions at $x = 0$ and $x = L$ are the same — u and $\partial u/\partial x$ both vanish. It may then be shown that equation (25.19) is replaced by

$$\cos \nu L \cosh \nu L = 1.$$

This still has an infinity of solutions, but the lowest frequency now corresponds to $\nu L \approx 4.73$, or

$$k \approx \frac{22.373c}{L^2}.$$

So for the same bar, clamping at both ends results in higher natural frequencies of oscillation.

So where *does* equation (25.15) come from? For the source of the second term, incorporating the fourth-order derivative with respect to x, return to equation (12.13) in Section 12, which is (to reiterate):

$$EI\frac{d^4y}{dx^4} = -f(x),$$

where E is Young's modulus, I is the moment of inertia and $f(x)$ — measured positive downwards, along with y — is the load per unit length.

In the present notation, replace y with u as the vertical beam displacement and also change from an ordinary derivative to a partial one, reflecting the fact that u depends on both x and t. Thus,

$$EI\frac{\partial^4 u}{\partial x^4} = -f(x).$$

The source of the right-hand load term can now be borrowed from Newton's force = mass × acceleration equation, which in the present context will be

$$\text{force} = \rho A dx \frac{\partial^2 u}{\partial t^2},$$

working on the basis of an element of mass being $\rho A dx$. But $f(x)$ is the force per unit length, so the dx term may be omitted, giving

$$EI\frac{\partial^4 u}{\partial x^4} = -\rho A\frac{\partial^2 u}{\partial t^2},$$

which, when rearranged, is the equation sought.

Chapter 26

CHAOTIC (BUT NOT RANDOM)

IN earlier sections, when describing and solving ordinary or partial differential equations, as well as difference equations, I was implicitly concentrating on those that were at least intuitively understandable, perhaps predictable and maybe even periodic — and also those which in general could be described as 'well-behaved' (for want of a better description). This chapter, in contrast, provides a glimpse of equations that are none of these things.

My dictionary defines the word chaos as 'formless void' or 'utter confusion', the latter conjuring up a picture of a ball of wool after a kitten has got at it. In mathematics, however, things are slightly more focused: there are various definitions [73], but 'aperiodic, unpredictable behaviour of deterministic nonlinear dynamical systems' is perhaps the most pertinent and the easiest to understand. And it is important to appreciate that the unpredictability has nothing to do with randomness: the output may appear random, but such behaviour has stemmed from entirely deterministic rules.

Nonlinearity is frequently a key attribute of such equations — although not all nonlinear equations exhibit chaotic behaviour. The previous sentence assumes that the word 'nonlinear' is self-explanatory; and certainly if the *dependent* variable in an equation occurs with powers other than one, or is embedded in more complicated functions such as exponentials, it is mostly safe to conclude that the equation is nonlinear. However, as will be seen below, it is also possible to obtain chaotic behaviour from equations that I would have confidently asserted were purely linear, and that would give predictable and maybe periodic output.

Continuing in the same vein, who would conceive that a particular geometric shape could enclose a finite planar area with a potentially infinite boundary? Yet such a structure is mathematically consistent, besides sharing some features with the chaotic examples with which we start this chapter.

The Lorenz[1] equations below are ideal for illustrating what we are initially

---

[1] Edward N. Lorenz, 1917 to 2008.

talking about here:

$$\frac{dx}{dt} = \sigma(y-x),$$
$$\frac{dy}{dt} = x(\rho - z) - y, \qquad (26.1)$$
$$\frac{dz}{dt} = xy - \beta z,$$

with constant values for σ, β and ρ; the nonlinearities appear here on the right-hand-sides as the products xz and xy. These equations were originally defined to form a simplified model of atmospheric convection, namely a two-dimensional fluid layer warmed from below and cooled from above. The underlying Navier-Stokes equations for fluid flow are nonlinear, coupled, partial differential equations, for which there are precious few exact analytic solutions. The above triplet, in contrast, is rather more amenable to investigation. Referring to [74], the quantity x is proportional to the intensity of the convective motion, y is proportional to the temperature difference between the ascending and descending currents, and z is proportional to the distortion of the vertical temperature profile from linearity.

In solving the above, we will stick with Lorenz's parameter values: $\sigma = 10$, $\beta = \frac{8}{3}$ and $\rho = 28$ (since the equations have now achieved a life and fame of their own, the exact meanings of the parameters are not especially relevant here). Solving the above triplet using the Runge-Kutta fourth-order scheme with a time step of 0.01 s, we end up with the familiar 'owl-face' diagram shown in Figure 26.1.

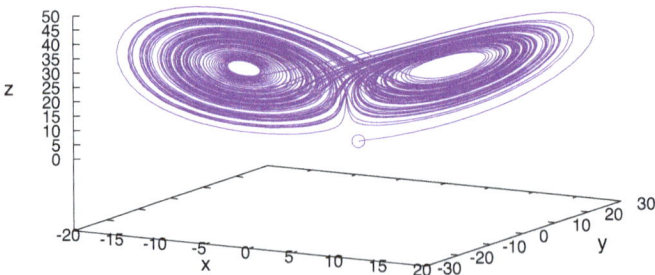

Figure 26.1: Lorenz equations output in three dimensions

Time is measured along the three-dimensional trajectory, with the small circle marking the starting point at $x = 1$, $y = 1$, $z = 1$. There is structure here, in that the trajectory tends to exhibit twin orbit-like behaviour, each orbit being broadly constrained in spatial terms, but unpredictably switching from one side to the other. The motion looks as if it should be periodic, but actually is not, as can be seen from the individual traces of x, y and z versus time:

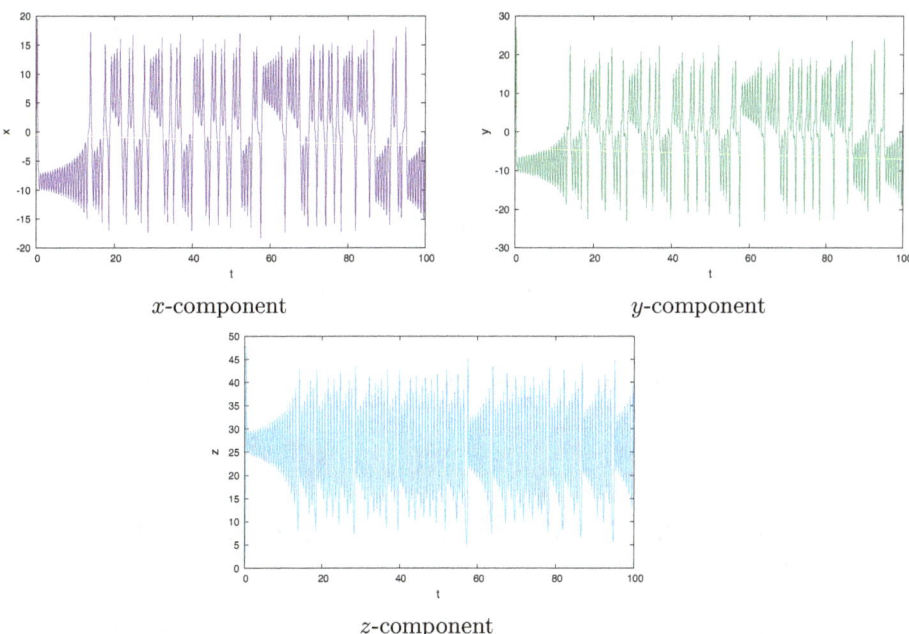

Figure 26.2: x, y and z versus time for the Lorenz equations

For each of these, following the initial jump into the left-hand 'eye' in Figure 26.1, the trajectory starts out periodic and with increasing amplitude, corresponding to spiralling outward. It then abruptly jumps across to the right-hand 'eye' and orbits that for a while in a similar manner. Subsequent jumps between these two orbital foci then continue unpredictably for as long as the simulation continues.

This is decidedly strange, if we think about it; there are no random contributory elements in the above triplet of equations, and yet the trajectory behaves quasi-stochastically. To put it another way, the observed erratic behaviour is entirely intrinsic in deterministic equations. And it is not as if this is an isolated instance — the same sort of thing occurs in areas as diverse as electronic circuits [75] (see below), planetary dynamics [76, 77], leaky taps [78], earth's climate [79], weather prediction [80], pandemics [81] and a whole lot more[2]. But before elaborating a bit on a couple of these areas, it is worth illustrating another aspect of chaos: sensitivity to initial conditions.

When solving differential equations, it is reasonable, is it not, to expect that small changes in the initial conditions will give rise to only small changes in the resulting behaviour over time? Not so for chaotic systems — even tiny variations at initiation can change the output considerably. This can be illustrated here by re-running the above system of equations (26.1) with $x = y = 1$, $z = 1 + 10^{-8}$. Plotting the difference in the resulting values of x over time gives Figure 26.3.

---

[2] Try an internet search with the phrase 'chaotic behaviour'.

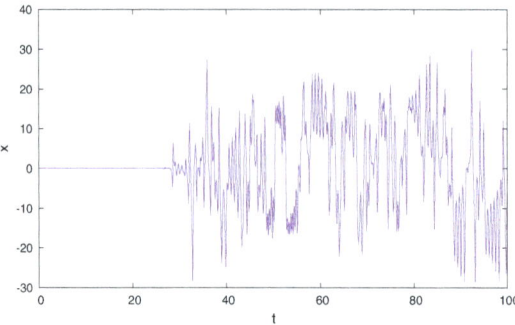

Figure 26.3: Difference in x vs t by starting with $z = 1 + 10^{-8}$ rather than $z = 1$

Up to about 30 s from the start, the change has minimal effect, but then there is a very abrupt deviation in x, with the same behaviour occurring in y and z. Although the three-dimensional 'owl-eye' picture resulting from this perturbation at the start is similar in character to Figure 26.1, it is different in numeric terms.

This sensitivity to initial conditions is one reason why weather forecasts come with a time-window of validity, perhaps 5 days or so into the future. Even if the initial weather state could be defined over a fantastically detailed spatial grid, the initial values cannot be set with infinite precision. And the governing partial differential equations cannot be solved exactly either. To some extent, the forecasts can be made more reliable by generating a set of 'candidate' weather pictures, each with slightly different settings. This is reminiscent of the Monte Carlo approach that is used with random systems, and motivated by similar considerations. Since chaotic systems exhibit what looks like random behaviour, it makes sense to borrow such concepts as mean and covariance from stochastic mathematics.

Go back now to Figure 26.2, and look at the initial behaviour of x, y and z, up to about the 15 second mark and just before the first orbital switch. Apart from the increasing orbital amplitude, this looks like stable periodic motion; so if one had only a short time to observe this system, the conclusion might well be that stability prevails. There are echoes here of planetary motion, in that stability can persist for millions of years (fortunately for life on earth), whereas the underlying behaviour is actually chaotic.

Perhaps of more pressing concern at the moment than planetary stability, though, is the question of climatic stability [79]. It is worrying enough to know that the climate can exhibit intrinsically chaotic behaviour, in the absence of any human interventions.

Surprisingly, not much is required in the way of nonlinearity to generate chaotic behaviour, as can be illustrated with a simplified form of Chua's electronic circuit [75]. The equations in this case are of the form:

$$C_1 \frac{dv_1}{dt} = G(v_2 - v_1) - g(v_1),$$
$$C_2 \frac{dv_2}{dt} = G(v_1 - v_2) + I, \quad (26.2)$$
$$L \frac{dI}{dt} = -v_2,$$

where v_1 and v_2 are voltages, I is current, C_1, C_2 are capacitances, L is inductance and G is resistance. The only nonlinear element in these equations is the function $g(.)$, which takes the piecewise-linear form shown in Figure 26.4.

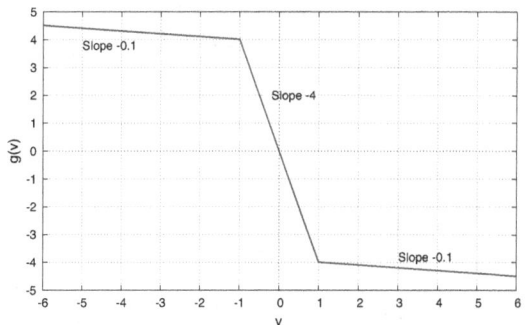

Figure 26.4: Nonlinear function $g(.)$ for the (simplified) Chua circuit

The corresponding circuit diagram is provided in Figure 26.5, with the nonlinear resistor on the right-hand side (this component can be constructed electronically, by the way, with details in [82]).

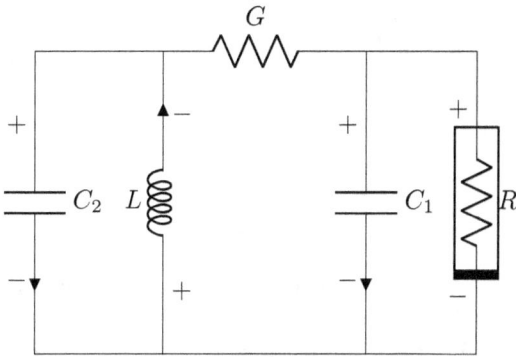

Figure 26.5: Corresponding chaotic circuit diagram

Going back to equation set 26.2, v_1 and v_2 are the voltages across capacitors C_1 and C_2 respectively, while I is the current through the inductance L. The values of the parameters have been set as follows:

$$\frac{1}{C_1} = 10, \quad \frac{1}{C_2} = 0.5, \quad \frac{1}{L} = 0.7 \text{ and } G = 0.7.$$

Starting the equations off arbitrarily with $v_1 = v_2 = I = 1$ and solving over 300 s with a time step of 0.02 s results in Figure 26.6. The small orange circle marks the starting point.

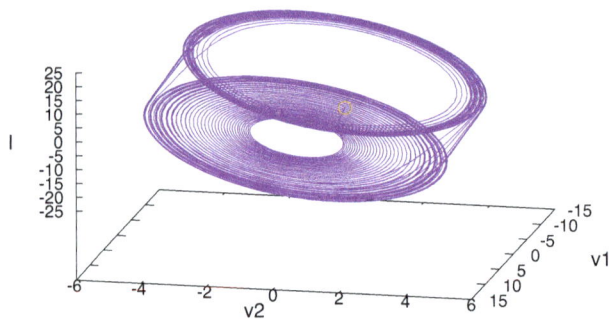

Figure 26.6: Output of the nonlinear Chua circuit

If the simulation is terminated prematurely at 100 s, the jump in the current I from the lower to the upper plane will be missed and the behaviour looks no more than periodic (albeit with increasing amplitude, similar to Figure 26.2). This can be seen more clearly if I is plotted against t, as in Figure 26.7.

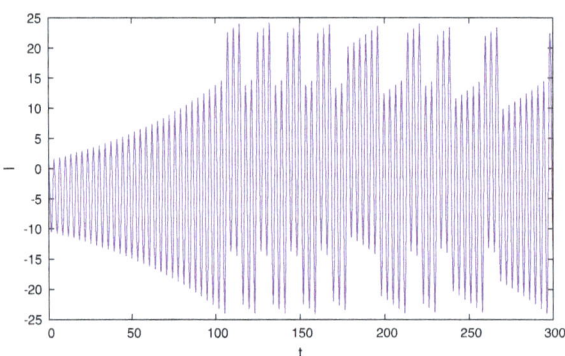

Figure 26.7: I versus t for the Chua circuit

Just as an experiment, I tried substituting a purely linear function for $g(.)$ in equation set 26.2, using $g(v_1) = -0.1v_1$. As might have been expected, this merely produced periodic output with growing amplitude, with no indication of unpredictable behaviour. This *linear* form of the function $g(.)$ also admits the use of the Laplace Transform solution method (Section 20.1), which is inapplicable to equation (26.2).

We now turn to a different manifestation of chaos, generated by an even simpler equation than in the above differential equations. This is the *logistic map*,

$$x_{k+1} = \mu x_k (1 - x_k),$$

where k stands for an iteration counter and μ is a parameter in the range zero to 4. The idea is to start things off with, say, $x_0 = 0.5$, choose a value of μ and then iterate over numerous values of k until x_k 'converges'. The last word here is put in quotes, since the convergence can be to a set of different values rather than the expected singleton, as will be seen from the graphical output below. Computationally, the process is simple: run over k up to $k = 300$, say, but output only the last 200 or so x_k values to plot on a graph, as shown in Figure 26.8.

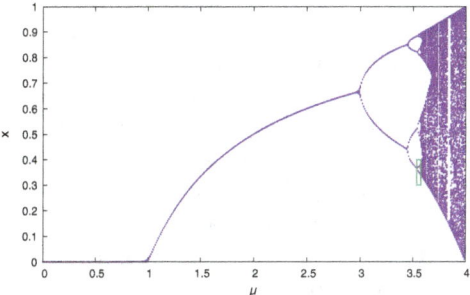

Figure 26.8: Plot of x_k versus μ for $\mu \in [0, 4)$

For μ up to about 1, x_k converges uniformly to zero. For μ between 1 and just less than 3, a single convergence value is still achieved, albeit monotonically growing with μ. Thereafter, things become more interesting with a bifurcation in values: x_k starts to alternate between two different points, each of which then bifurcates further once μ reaches 3.4 or so. As μ increases further, these bifurcations split again and again until fully chaotic behaviour occurs at about $\mu = 3.57$. Divergence occurs for values of $\mu > 4$.

Zooming in on the green box area in Figure 26.8 gives Figure 26.9.

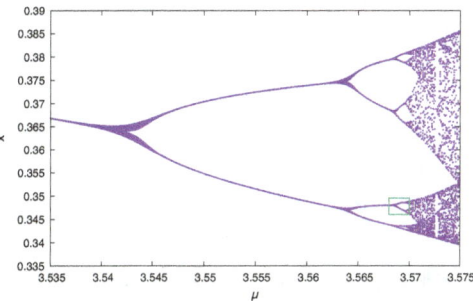

Figure 26.9: Plot of x_k versus μ for $\mu \in [3.535, 3.575]$

And if the green box in this one is expanded, we end up with Figure 26.10.

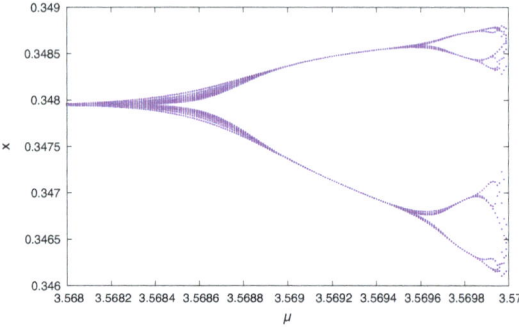

Figure 26.10: Plot of x_k versus μ for $\mu \in [3.568, 3.57]$

The bifurcation process continues at least two steps further before chaos intervenes. This sort of self-symmetry is characteristic of fractal behaviour, in which the magnified structure mirrors that at earlier levels ([73]).

Fractal behaviour can also be constructed as well as discovered, and a good example of a fractal curve is the Koch[3] 'snowflake', which is particularly simple to generate (as well as being fun to program). This is built up using an equilateral triangle as a basic structure, and then adding progressively smaller equilateral triangular outcrops on to the straight sections, as shown in Figure 26.11 (easier to illustrate than to describe).

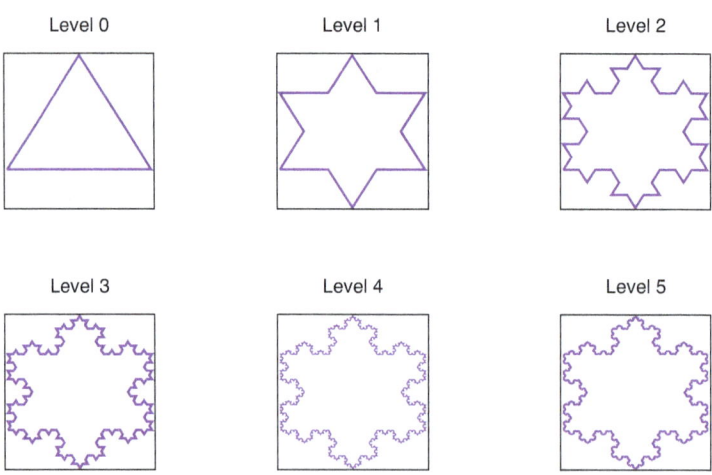

Figure 26.11: Five iterations of the Koch curve

This fractal curve has the paradoxical property that the area inside the curve is

---
[3]Nils von Koch, 1870 to 1924.

strictly finite (as can be inferred from the boxes drawn around each iteration, but also worked out below), while the curve length grows with each step[4]. So the length is in principle infinite if the number of iterations is infinite, which I admit is not obvious from looking at Figure 26.11. It can be reasoned out with the following logic.

Suppose that the length of the sides in the original triangle (level 0 in Figure 26.11) is L_0. Then concentrate on just the base of the level 1 curve in Figure 26.11, and annotate as shown in Figure 26.12.

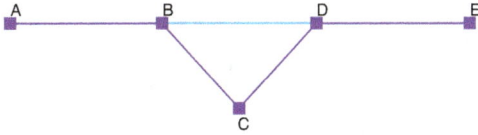

Figure 26.12: Base of the level 1 curve

This divides the original line of length L_0 into the three pieces AB, BD and DE, each of which will be of length $L_0/3$. Cut out the light blue line segment BD shown, paste it in to form the new segment BC and add another segment CD to join up. The length of the level 1 line will now be $4L_0/3$. Or, using an obvious notation,

$$L_1 = \frac{4}{3} L_0.$$

Now imagine doing exactly the same operation on each of the individual segments AB, BC, CD and DE in the level 1 curve, in order to create the level 2 curve. Each such segment will have length $L_1/4$, so splitting it into 3 and replacing the centre section with two sides of an equilateral triangle will give a new segment length as $(4/3) \times (L_1/4) = L_1/3$. The total length of the level 2 curve will then be given by

$$L_2 = \frac{4}{3} L_1.$$

And so on. For k iterations, we can conclude that

$$L_k = \frac{4}{3} L_{k-1} = L_0 \left(\frac{4}{3}\right)^k,$$

and then multiply by 3 to get the length of the whole curve. The length grows inexorably with each iteration.

Working out the area inside the curve is not that difficult either. Two insights are helpful here: when adding a triangle on to each straight side, it is one-third smaller in size[5] than at the preceding level (since each straight side is subdivided

---

[4] This is reminiscent of claims that coastlines are infinite, in that more and more detailed maps reveal greater and greater lengths.
[5] Meaning length of side.

into three). This means that its triangular area must be one-ninth the size of its predecessor. Also, each level contains 4 more triangles than are at the preceding level. So we can build up the following table, concentrating for now on just one side of the level 0 triangle:

Level	Number of triangles	Area relative to A_0
1	1	$\frac{1}{9}$
2	4	$\left(\frac{1}{9}\right)^2$
3	4^2	$\left(\frac{1}{9}\right)^3$
4	4^3	$\left(\frac{1}{9}\right)^4$
and so on.		

The quantity A_0 is the area of the level 0 triangle, which is given by $L_0^2 \sqrt{3}/4$, using the above-mentioned length quantity. Adding up the various contributions, we find that the *additional* area S_k resulting from k levels above zero (and so additive to A_0) will be given by:

$$S_k = 3A_0 \left[\left(\frac{1}{9}\right) + 4\left(\frac{1}{9}\right)^2 + 4^2\left(\frac{1}{9}\right)^3 + 4^3\left(\frac{1}{9}\right)^4 + \ldots + 4^{k-1}\left(\frac{1}{9}\right)^k \right],$$

$$= \frac{A_0}{3}\left[1 + \left(\frac{4}{9}\right) + \left(\frac{4}{9}\right)^2 + \left(\frac{4}{9}\right)^3 + \ldots + \left(\frac{4}{9}\right)^{k-1}\right].$$

The factor 3 out front in the first equality just reflects the fact that the A_0 triangle has 3 sides. The second equation can also be written as:

$$S_k = \frac{A_0}{3}\left[\frac{1 - (4/9)^k}{1 - 4/9}\right] \quad \text{(see Section 4.1)},$$

$$= \frac{3A_0}{5}\left\{1 - \left(\frac{4}{9}\right)^k\right\}.$$

This is the *additional* area resulting from all those extra triangles, so the total area inside a level k Koch curve must be

$$A_k = A_0\left[1 + \frac{3}{5}\left\{1 - \left(\frac{4}{9}\right)^k\right\}\right],$$

$$\to \frac{8}{5}A_0 \text{ as } k \to \infty.$$

The potentially infinite boundary of the Koch curve combined with its finite enclosed area provides an added constraint on the applicability of equation (13.1) in Section 13.1. However, provided k remains finite, it is expected that equation (13.1) will remain valid (an exercise for the reader).

Fractal structures can also be created in both lower and higher spatial dimensions [73], and a one-dimensional example is known as *Cantor*[6] *dust*, illustrated in Figure 26.13.

---

[6]Georg Cantor, 1845 to 1918.

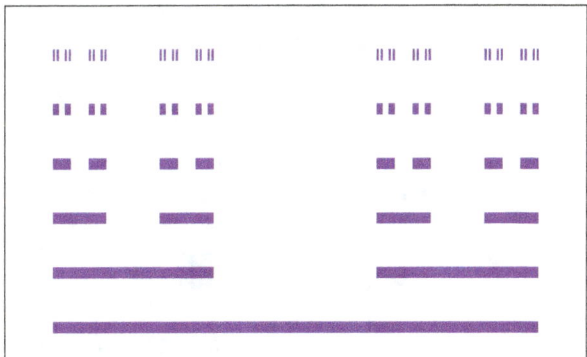

Figure 26.13: Cantor dust

This created by taking a line of unit length (shown here as a bar for visibility), chopping out the middle one-third, and then repeating the process on each subline ad infinitum. One ends up with an infinite scattering of points having zero total length. It turns out that this structure is a good model for errors occurring in transmission lines.

Chapter 27

TSUNAMI WAVES

EVEN a brief look at [83] shows how often tsunamis have occurred throughout history, whether triggered by earthquakes, volcanic eruptions (probably the most common source) or landslides (which can occur above or below water), or a combination of all three. Out at sea, the tsunami wave appears benign enough, but the real damage occurs at landfall when the approach into shallower water causes the wave amplitude to rear up into terrifying heights, threatening considerable loss of life and extensive damage even far inland.

Several of the earth's oceans now have early-warning systems in place to allow for the evacuation of coastal populations at risk. For this to be effective, once the initial disturbance (such as an earthquake) is detected, it is necessary to be able to predict when the wave will hit the surrounding coastlines. This, in turn, requires estimates of speed of travel.

It may seem paradoxical that the shallow-water equations are suitable for this analysis, but tsunami wavelengths can be a hundred kilometres or more, while ocean depths are at most around 4 km. So from the tsunami's perspective, the water it travels through is shallow (or even very shallow).

The equations for waves on shallow water are as follows [84]:

$$\frac{\partial u}{\partial t} + u\frac{\partial u}{\partial x} + v\frac{\partial u}{\partial y} + g\frac{\partial \eta}{\partial x} = 0,$$
$$\frac{\partial v}{\partial t} + u\frac{\partial v}{\partial x} + v\frac{\partial v}{\partial y} + g\frac{\partial \eta}{\partial y} = 0,$$
$$\frac{\partial \eta}{\partial t} + \frac{\partial}{\partial x}\left[(\eta + h)u\right] + \frac{\partial}{\partial y}\left[(\eta + h)v\right] = 0,$$

in which the coordinate x and its companion velocity component are horizontal and from left to right; y and v are also horizontal and into the page, while the vertical z motion is neglected (indeed, a process of depth-averaging in the derivation of these equations has removed the vertical z coordinate and its associated velocity component).

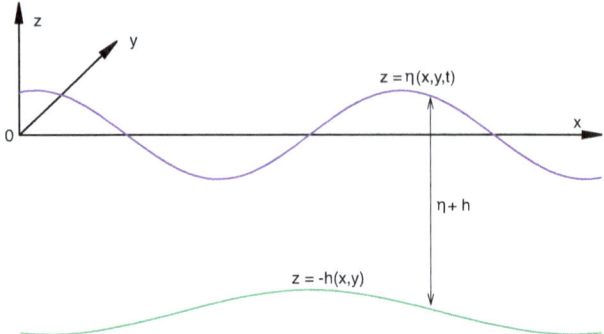

Figure 27.1: Shallow water coordinates

The quantity $\eta(x, y, t)$ defines the shape of the free surface and $h(x, y)$ describes the sea floor topography (which need not be flat), while the parameter g is the local acceleration due to gravity. The coordinate z is measured relative to the average sea surface level, so h is actually negative, with its sign already incorporated into the above equations; see Figure 27.1.

For those readers interested, the derivation of the above equations is outlined in Appendix L.

Suppose now that u, v and η are all small quantities, so that we can obtain linearised forms. That is,

$$u = \epsilon u' + O(\epsilon^2), \quad v = \epsilon v' + O(\epsilon^2), \quad \eta = \epsilon \eta' + O(\epsilon^2),$$

where ϵ is a small quantity; the dash notation is just a discriminant here and does not indicate derivative. To first-order in ϵ, and ignoring the terms of $O(\epsilon^2)$ and higher, the above equations become:

$$\frac{\partial u'}{\partial t} + g \frac{\partial \eta'}{\partial x} = 0,$$

$$\frac{\partial v'}{\partial t} + g \frac{\partial \eta'}{\partial y} = 0,$$

$$\frac{\partial \eta'}{\partial t} + \frac{\partial}{\partial x}(hu') + \frac{\partial}{\partial y}(hv') = 0.$$

These can actually be solved analytically, at least so far as η is concerned, and the various steps involved are set out in some detail below. Begin by multiplying the first two of these by $\sqrt{h}$, noting that h is independent of time:

$$\frac{\partial}{\partial t}\left(u'\sqrt{h}\right) + g\sqrt{h}\,\frac{\partial \eta'}{\partial x} = 0,$$

$$\frac{\partial}{\partial t}\left(v'\sqrt{h}\right) + g\sqrt{h}\,\frac{\partial \eta'}{\partial y} = 0,$$

$$\frac{\partial \eta'}{\partial t} + \frac{\partial}{\partial x}(hu') + \frac{\partial}{\partial y}(hv') = 0.$$

Next multiply the last of these by $\sqrt{g}$, noting that g is a constant:

$$\frac{\partial}{\partial t}\left(u'\sqrt{h}\right) + g\sqrt{h}\,\frac{\partial \eta'}{\partial x} = 0,$$

$$\frac{\partial}{\partial t}\left(v'\sqrt{h}\right) + g\sqrt{h}\,\frac{\partial \eta'}{\partial y} = 0,$$

$$\frac{\partial}{\partial t}(\eta'\sqrt{g}) + \frac{\partial}{\partial x}(hu'\sqrt{g}) + \frac{\partial}{\partial y}(hv'\sqrt{g}) = 0.$$

By writing the unitary g and h quantities as products of square roots, these can be further written in the form:

$$\frac{\partial}{\partial t}\left(u'\sqrt{h}\right) + \sqrt{gh}\,\frac{\partial}{\partial x}(\eta'\sqrt{g}) = 0,$$

$$\frac{\partial}{\partial t}\left(v'\sqrt{h}\right) + \sqrt{gh}\,\frac{\partial}{\partial y}(\eta'\sqrt{g}) = 0,$$

$$\frac{\partial}{\partial t}(\eta'\sqrt{g}) + \frac{\partial}{\partial x}\left(u'\sqrt{h}\sqrt{gh}\right) + \frac{\partial}{\partial y}\left(v'\sqrt{h}\sqrt{gh}\right) = 0.$$

The above sequence resulting in this new triplet of equations may not seem to have gained anything, but the structure is clearer if the following new variables are introduced:

$$A(x,y,t) = u'\sqrt{h},\ \ B(x,y,t) = v'\sqrt{h},\ \ C(x,y,t) = \eta'\sqrt{g}\ \text{and}\ \kappa(x,y) = \sqrt{gh}.$$

That is,

$$\frac{\partial A}{\partial t} + \kappa\frac{\partial C}{\partial x} = 0,$$

$$\frac{\partial B}{\partial t} + \kappa\frac{\partial C}{\partial y} = 0,$$

$$\frac{\partial C}{\partial t} + \frac{\partial}{\partial x}(\kappa A) + \frac{\partial}{\partial y}(\kappa B) = 0.$$

The aim now is to use the first two equations here to eliminate A and B from the last equation, and so end up with a single second-order equation in C. To this end, multiply the first two equations by κ (recalling that this is independent of time) and rearrange. In addition, take the partial derivative of the last equation with respect to time:

$$\frac{\partial}{\partial t}(\kappa A) = -\kappa^2\frac{\partial C}{\partial x},$$

$$\frac{\partial}{\partial t}(\kappa B) = -\kappa^2\frac{\partial C}{\partial y},$$

$$\frac{\partial^2 C}{\partial t^2} + \frac{\partial^2}{\partial x\partial t}(\kappa A) + \frac{\partial^2}{\partial y\partial t}(\kappa B) = 0.$$

The final step is to take the partial derivatives of the first and second equations with respect to x and y respectively:

$$\frac{\partial^2}{\partial x \partial t}(\kappa A) = -\frac{\partial}{\partial x}\left(\kappa^2 \frac{\partial C}{\partial x}\right),$$

$$\frac{\partial^2}{\partial y \partial t}(\kappa B) = -\frac{\partial}{\partial y}\left(\kappa^2 \frac{\partial C}{\partial y}\right),$$

$$\frac{\partial^2 C}{\partial t^2} + \frac{\partial^2}{\partial x \partial t}(\kappa A) + \frac{\partial^2}{\partial y \partial t}(\kappa B) = 0.$$

The first two equations can now be substituted into the last, ending up with the following:

$$\frac{\partial^2 C}{\partial t^2} = \frac{\partial}{\partial x}\left(\kappa^2 \frac{\partial C}{\partial x}\right) + \frac{\partial}{\partial y}\left(\kappa^2 \frac{\partial C}{\partial y}\right).$$

The basic structure of the answer we want can be obtained without explicitly solving this for C: suppose that C is a function of x and t only, and that κ is a constant (or, at least, is sufficiently slowly-varying with x so that its derivative can be neglected). Therefore,

$$\frac{\partial^2 C}{\partial t^2} = \kappa^2 \frac{\partial^2 C}{\partial x^2},$$

which is the wave equation in x and t, with general solution of the form

$$C = f(x \pm \kappa t).$$

So if the disturbance was initially sinusoidal in space, we expect that it will propagate in space and time according to

$$C \propto \sin(x \pm \kappa t),$$

in which κ is the speed of propagation. We thus expect the tsunami disturbance propagation speed to be

$$V = \sqrt{gh}.$$

This is independent of the disturbance wavelength and dependent only on the water depth (plus gravity). The same result can be obtained using a different approach [38], which does not need to be replicated here.

Analysing the behaviour of the wave at landfall requires more complex (although still approximate) equations than have been examined here, most notably in retaining the nonlinear terms. This need for different approximations is a consequence of the somewhat fragmentary nature of fluid dynamics: the foundational Navier-Stokes equations are too complicated to solve analytically except in simple situations, which has given rise to something of a rabbit-warren of interlinked 'special cases', each with its own set of approximations and domain of applicability. Only in fairly recent years has it been possible to solve the main equations themselves using numerical methods.

Chapter 28

SPECIAL RELATIVITY

THE year 1905, when Einstein[1] published his landmark paper on special relativity [85], marked the beginning of several decades of upheaval in physics, when long-established views on the nature of space, time, matter and energy needed to be revised. There had been rumblings of inconsistencies prior to that, for instance concerning the lack of evidence for any substance acting as a medium of transmission for electromagnetic radiation, or the peculiar nature of black-body radiation, but Einstein's paper started the process of clarification.

The equations governing electromagnetic radiation are a good place for us to begin. Considering only one dimension of space, the partial differential equation for radiation ϕ in a vacuum is, from Section 25.2,

$$\frac{\partial^2 \phi}{\partial t^2} = c^2 \frac{\partial^2 \phi}{\partial x^2}, \qquad (28.1)$$

where t denotes time, x is the spatial dimension and c is the speed of light.

Suppose now that we have two observers, one of which is deemed static and the other moving with a constant speed v relative to the first. The relative motion is entirely on the x axis, so the other two space coordinates can be ignored. Following Newton, the relation between the two coordinate systems was assumed to be as follows:

$$x' = x - vt, \qquad (28.2)$$
$$t' = t, \qquad (28.3)$$

where x, t are used in the 'static' coordinate system and x', t' in the 'moving' system[2]. The superscript dash here is just used as a discriminant, and does not denote differentiation. Regarding the equation $t' = t$, there was in fact no discrimination made between the two times: time was expected to be universal and the same for everyone, whether moving or not.

---

[1] Albert Einstein, 1879 to 1955.
[2] The quotation marks are used here because it is an entirely arbitrary matter as to which frame of reference is moving and which is not.

But these coordinate transformations do not preserve the form of equation (28.1); this can be seen by deriving the derivative operators, as follows:

$$\frac{\partial}{\partial x} = \frac{\partial x'}{\partial x}\frac{\partial}{\partial x'} + \frac{\partial t'}{\partial x}\frac{\partial}{\partial t'} = \frac{\partial}{\partial x'},$$

$$\frac{\partial}{\partial t} = \frac{\partial x'}{\partial t}\frac{\partial}{\partial x'} + \frac{\partial t'}{\partial t}\frac{\partial}{\partial t'} = -v\frac{\partial}{\partial x'} + \frac{\partial}{\partial t'}.$$

Substitute these into equation (28.1):

$$\left(-v\frac{\partial}{\partial x'} + \frac{\partial}{\partial t'}\right)\left(-v\frac{\partial}{\partial x'} + \frac{\partial}{\partial t'}\right)\phi = c^2\left(\frac{\partial}{\partial x'}\right)\left(\frac{\partial}{\partial x'}\right)\phi,$$

which, after some simplification and rearrangement, becomes:

$$\frac{\partial^2\phi}{\partial t'^2} = (c^2 - v^2)\frac{\partial^2\phi}{\partial x'^2} + 2v\frac{\partial^2\phi}{\partial x'\partial t'}.$$

This is of a completely different form to equation (28.1). In particular, it predicts that the speed of light transmission in the x', t' frame will be given by $\sqrt{c^2 - v^2}$, and this could even be zero if $v = c$ (and undefined if $v > c$). In contrast, numerous meticulous experiments had failed to find any variation in the speed of light, regardless of how observers moved relative to one another.

It was already known, however, that equation (28.1) remains invariant under the Lorentz transformations[3] [86]:

$$x' = \frac{x - vt}{\sqrt{1 - v^2/c^2}}, \tag{28.4}$$

$$t' = \frac{t - vx/c^2}{\sqrt{1 - v^2/c^2}}. \tag{28.5}$$

This can be seen by again deriving the relevant derivative operators:

$$\frac{\partial}{\partial x} = \frac{1}{\gamma}\left(\frac{\partial}{\partial x'} - \frac{v}{c^2}\frac{\partial}{\partial t'}\right),$$

$$\frac{\partial}{\partial t} = \frac{1}{\gamma}\left(\frac{\partial}{\partial t'} - v\frac{\partial}{\partial x'}\right),$$

writing $\gamma = \sqrt{1 - v^2/c^2}$ for convenience. Therefore,

$$\frac{\partial^2}{\partial x^2} = \frac{1}{\gamma^2}\left(\frac{\partial^2}{\partial x'^2} - \frac{2v}{c^2}\frac{\partial^2}{\partial x'\partial t'} + \frac{v^2}{c^4}\frac{\partial^2}{\partial t'^2}\right),$$

$$\frac{\partial^2}{\partial t^2} = \frac{1}{\gamma^2}\left(\frac{\partial^2}{\partial t'^2} - 2v\frac{\partial^2}{\partial x'\partial t'} + v^2\frac{\partial^2}{\partial x'^2}\right).$$

Substitute these into equation (28.1) to get:

$$\frac{1}{\gamma^2}\left(\frac{\partial^2\phi}{\partial t'^2} - 2v\frac{\partial^2\phi}{\partial x'\partial t'} + v^2\frac{\partial^2\phi}{\partial x'^2}\right) = \frac{c^2}{\gamma^2}\left(\frac{\partial^2\phi}{\partial x'^2} - \frac{2v}{c^2}\frac{\partial^2\phi}{\partial x'\partial t'} + \frac{v^2}{c^4}\frac{\partial^2\phi}{\partial t'^2}\right).$$

---

[3]Hendrik Lorentz, 1853 to 1928.

The derivative cross-term $\partial^2\phi/\partial x'\partial t'$ cancels out and the remaining terms may then be rearranged to give:

$$\frac{1}{\gamma^2}\left(1-\frac{v^2}{c^2}\right)\frac{\partial^2\phi}{\partial t'^2} = \frac{1}{\gamma^2}\left(c^2-v^2\right)\frac{\partial^2\phi}{\partial x'^2},$$

$$= \frac{c^2}{\gamma^2}\left(1-\frac{v^2}{c^2}\right)\frac{\partial^2\phi}{\partial x'^2},$$

which is exactly equivalent to the electromagnetic equation (28.1) in the x', t' frame. So the same equation applies and the same physics is expected.

Einstein asserted that the speed of light must be an invariant, regardless of observer movement, and that equations (28.4) and (28.5) were the correct ones to use when transforming between frames of reference that are moving at a constant speed v relative to one another. These transformation equations then have observable consequences, most notably length contraction and time dilation.

To see this, clarify the nomenclature by calling the undashed 'static' frame of reference with coordinates x, t as frame A and the dashed 'moving' frame with coordinates x', t' as B. Frame B is assumed to be moving to the right relative to frame A at a speed v. Then lay out a ruler in B with end-points x'_1, x'_2 seen at the same instant t by A. Then from equation (28.4),

$$x'_1 = \frac{x_1 - vt}{\sqrt{1-v^2/c^2}} \quad \text{and} \quad x'_2 = \frac{x_2 - vt}{\sqrt{1-v^2/c^2}},$$

The difference of these will then be:

$$L' = x'_2 - x'_1 = \frac{x_2 - x_1}{\sqrt{1-v^2/c^2}} = \frac{L}{\sqrt{1-v^2/c^2}}.$$

Turning this around,

$$L = x_2 - x_1 = L'\sqrt{1-v^2/c^2},$$

and provided $v < c$, observer A sees the ruler as having a shorter length than is measured by B: this is termed *Fitzgerald*[4] *contraction*. This does not mean that B has chopped bits off the local ruler — from his or her point of view, nothing has changed. And the situation is entirely symmetric: B sees A's ruler as being shorter. This has to be so, for otherwise there would be a preferred frame of reference — and there isn't. We can confirm the symmetry mathematically by examining the inverse Lorentz transformations, which are:

$$x = \frac{x' + vt'}{\sqrt{1-v^2/c^2}}, \qquad (28.6)$$

$$t = \frac{t' + vx'/c^2}{\sqrt{1-v^2/c^2}}, \qquad (28.7)$$

---

[4] George Fitzgerald, 1851 to 1901.

and these have the same form as equations (28.4) and (28.5), but with a reversed sign on v (since B moving to the right relative to A is the same as A moving to the left relative to B).

Now suppose that B arranges for light pulses to be emitted at times t'_1 and t'_2 at the same point $x' = 0$. Then, from equation (28.7), we have:

$$\Delta t = t_2 - t_1 = \frac{t'_2 - t'_1}{\sqrt{1 - v^2/c^2}},$$

$$= \frac{\Delta t'}{\sqrt{1 - v^2/c^2}}.$$

Adding some numbers into this, suppose $\Delta t' = 1$ hour and $v = 0.95c$; then Δt will be approximately 3.2 hours. So A sees the time difference between the two pulses as *longer* than does B, equivalent to seeing a reduced clock rate. This is called the time dilation effect, and occurs in a related context in general relativity (see Section 29.1).

To put the symmetry in the differences rather more informally, it is always the other fellow who has the short rulers and slow clocks[5].

The Lorentz transformations can also be used to work out the change in frequency (or wavelength) of radiation emitted in one frame and received in another, where there is relative motion at a constant speed between the two. Assume that a sinusoidal signal is emitted with angular frequency ω_A from frame A, of the functional form

$$\phi_A(x, t) = \sin \omega_A (t - x/c),$$

which satisfies equation (28.1). What does this signal look like when it is received at frame B? To find out, concentrate on the $\omega_A (t - x/c)$ quantity and substitute in the Lorentz transformations (28.6) and (28.7):

$$\omega_A (t - x/c) = \omega_A \left\{ \frac{t' + vx'/c^2}{\sqrt{1 - v^2/c^2}} - \frac{1}{c} \left(\frac{x' + vt'}{\sqrt{1 - v^2/c^2}} \right) \right\},$$

$$= \frac{\omega_A}{\sqrt{1 - v^2/c^2}} \left\{ t' \left(1 - \frac{v}{c}\right) - \frac{x'}{c} \left(1 - \frac{v}{c}\right) \right\},$$

$$= \omega_A \sqrt{\frac{1 - v/c}{1 + v/c}} \left(t' - \frac{x'}{c} \right).$$

So frame B receives a signal of the form

$$\phi_B(x', t') = \sin \omega_B (t' - x'/c),$$

where

$$\omega_B = \omega_A \sqrt{\frac{1 - v/c}{1 + v/c}}.$$

---

[5] At least in the absence of accelerations or gravitational fields.

If, as assumed here, v is positive and consistent with the frames moving apart, $\omega_B < \omega_A$ and the signal received at B is red-shifted[6] (moved toward the red end of the spectrum, with a lower frequency). On the other hand, if the frames are moving together, the signal will be blue-shifted. And the situation is symmetric: it doesn't matter which frame is considered to be 'static'.

These consequences of the Lorentz transformations have been confirmed by numerous detailed experiments, but they are admittedly counter-intuitive. We are so used to our concept of time being universal and inflexible that mentally coping with time dilation effects is quite a challenge; certainly I find it so. In defence of the earlier physics, it has to be said that these effects do not become apparent until v reaches some significant proportion of light speed; for $v/c \ll 1$, the familiar Galilean[7] or Newtonian transformations in equations (28.2) and (28.3) are retrieved.

For reasons that will become apparent, $v = c$ cannot be attained (except by photons), while particles with $v > c$ remain hypothetical. To justify the former statement, we can ask what happens if we send out a spaceship and arrange for it to accelerate at a constant rate along the x-axis for as long as we choose; this situation is examined a bit further in Section 28.1 below.

It can be seen from the above Lorentz transformation equations that space and time are now thoroughly intermixed — we cannot consider time and space as independent of one another. This leads to a redefinition of total Pythagorean lengths, replacing the familiar infinitesimal total spatial length,

$$ds^2 = dx^2 + dy^2 + dz^2,$$

by the more general equation:

$$ds^2 = c^2 dt^2 - \left(dx^2 + dy^2 + dz^2\right). \tag{28.8}$$

This is the equivalent of Pythagoras' theorem, but now in four-dimensional space time. It will be seen that the time coordinate dt is multiplied by c (which is needed for its contribution to have the dimensions of length) and that it has the opposite sign to the spatial contributions[8]. This difference reflects the fact that, from experience, we are able to move freely backward and forward in space, but in only one direction in time (always getting older ...).

The qualitative difference between time and space could have been predicted by going back to equation (28.1) for electromagnetic radiation. In three spatial dimensions, it is:

$$\frac{\partial^2 \phi}{\partial t^2} = c^2 \left(\frac{\partial^2 \phi}{\partial x^2} + \frac{\partial^2 \phi}{\partial x^2} + \frac{\partial^2 \phi}{\partial x^2}\right),$$

with time on the left and the three space components on the right. To put time on the same footing as space, replace t with $\xi = ict$, in which $i^2 = -1$ (see

---

[6] A more detailed analysis shows that there is signal amplitude attenuation of the same form, since the electric and magnetic field strengths also undergo Lorentz-type transformations [86].

[7] After Galileo Galilei, 1564 to 1642.

[8] An alternative convention may also be used, in which the spatial components are positive and the time contribution negative, but there are no physical consequences resulting from the change.

Chapter 17). With this substitution, it is found that

$$\frac{\partial^2 \phi}{\partial \xi^2} + \frac{\partial^2 \phi}{\partial x^2} + \frac{\partial^2 \phi}{\partial x^2} + \frac{\partial^2 \phi}{\partial x^2} = 0,$$

and the corresponding metric would be

$$-ds^2 = d\xi^2 + dx^2 + dy^2 + dz^2.$$

To be consistent, therefore, a point in the corresponding four-dimensional space-time is given by the vector:

$$\mathbf{X} = (ict, x, y, z). \qquad (28.9)$$

See Chapter 14 for a bit of discussion on vectors, but all we are doing here is attaching labels to the four coordinates in a consistent manner.

A different handle on the time dilation effect can be obtained from the definition of *proper time* τ, given by

$$d\tau = \frac{ds}{c}.$$

This is the time interval between events in a frame for which the events occur at the same spatial point. From the equation (28.8), with $dx = dy = dz = 0$, it is found that $dt = ds/c$ so it makes sense to generalise this to other geometries with different definitions of ds. The concept of proper time recurs in a related context in general relativity (see Chapter 29).

Now revert to equation (28.8) and set $x' = x - vt$ with y and z unchanged. Therefore, $dx' = dx - vdt$ and it is found that

$$\begin{aligned} ds^2 &= c^2 dt^2 - \left(dx'^2 + dy^2 + dz^2\right), \\ &= c^2 dt^2 - (dx + vdt)^2 - \left(dy^2 + dz^2\right), \\ &= \left(c^2 - v^2\right) dt^2 - \left(dx'^2 + 2v dx' dt + dy^2 + dz^2\right), \\ &= \left(c^2 - v^2\right) dt^2, \text{ if } dx' = dy = dz = 0. \end{aligned}$$

On this basis, the appropriate proper time $d\tau$ will be given by

$$d\tau = \frac{ds}{c} = \sqrt{1 - \frac{v^2}{c^2}} \, dt. \qquad (28.10)$$

This $d\tau$ will be the time interval measured in the 'moving' frame B, and will be shorter than the time interval dt measured in A. We thus retrieve the same time dilation effect via a different route.

The concept of proper time is needed if consistent Lorentz-type transformations for the velocities between different frames are to be obtained, although these specific conversions are not needed here. Of more relevance is to come up with a generalisation of the usual three-dimensional velocity vector (see Chapter 14):

$$\mathbf{v} = \left(\frac{dx}{dt}, \frac{dy}{dt}, \frac{dz}{dt}\right).$$

A suitable choice for a four-vector $\mathbf{V}$ is then obtained by differentiating the four-vector $\mathbf{X}$ in equation (28.9) with respect to the proper time τ:

$$\begin{aligned}\mathbf{V} &= \frac{d\mathbf{X}}{d\tau} = \frac{1}{\sqrt{1-v^2/c^2}} \frac{d\mathbf{V}}{dt}, \\ &= \frac{1}{\sqrt{1-v^2/c^2}} \left(ic, \frac{dx}{dt}, \frac{dy}{dt}, \frac{dz}{dt}\right), \\ &= \frac{1}{\sqrt{1-v^2/c^2}} (ic, \mathbf{v}),\end{aligned} \quad (28.11)$$

here replacing the last three components by the familiar velocity three-vector $\mathbf{v}$. With a consistent definition of four-velocity, it is possible to move on to define a similarly consistent momentum, which will also be a vector. In Newtonian dynamics, momentum $\mathbf{p}$ is defined as mass × velocity, or

$$\mathbf{p} = m\mathbf{v},$$

where m stands for the quantity usually understood to be the object's mass. This quantity $\mathbf{p}$ is then conserved during particle interactions, such as collisions. Mapping this concept over to four dimensional space-time leads us to look for conservation of a new quantity

$$\mathbf{P} = M\mathbf{V},$$

where M is 'mass-like' and needs to be determined. From the above definition of $\mathbf{V}$, we have

$$\mathbf{P} = \frac{M}{\sqrt{1-v^2/c^2}}(ic, \mathbf{v}) = m(ic, \mathbf{v}), \quad (28.12)$$

here defining

$$m = \frac{M}{\sqrt{1-v^2/c^2}}.$$

This is a vector equation for $\mathbf{P}$, so each component must be conserved, which means in turn that m itself is conserved and $m\mathbf{v}$ is conserved. The latter quantity, in particular, needs to be consistent with the 'classical limit' in which $v/c \ll 1$, meaning that M must be the same as the familiar object mass; it is usually termed m_0, or the object's *rest mass*.

We therefore get

$$m = \frac{m_0}{\sqrt{1-v^2/c^2}}. \quad (28.13)$$

So perceived mass, along with length and time, now also depends on speed. In particular, as $v \to c$, m becomes unboundedly large, meaning that its inertia (resistance to change in kinematic condition) grows without limit and infinite quantities of energy would be needed to push any object to light speed. (Photons — 'particles' of light — have zero rest mass and are exempt from this constraint).

The final step in this section is to gain some understanding as to where Einstein's famous equation $E = mc^2$ comes from. We have already needed to generalise the concept of momentum into four-vector form, and the same is needed for force. In Newtonian mechanics force is defined as the rate of change of momentum, so:

$$\mathbf{f} = \frac{d}{dt}(m\mathbf{v}) = \frac{d\mathbf{p}}{dt}.$$

In special relativity, it is then natural to move to the more general form in terms of proper time, and define:

$$\mathbf{F} = \frac{d\mathbf{P}}{d\tau} = \frac{d}{d\tau}\left\{m\left(ic, \mathbf{v}\right)\right\}, \text{ from equation (28.12),}$$

$$= \frac{dt}{d\tau}\frac{d}{dt}\left\{m\left(ic, \mathbf{v}\right)\right\}, \text{ changing the derivative from } \tau \text{ to } t,$$

$$= \frac{1}{\sqrt{1-v^2/c^2}}\left\{ic\frac{dm}{dt}, \frac{d}{dt}(m\mathbf{v})\right\}, \text{ from equation (28.10),}$$

$$= \frac{1}{\sqrt{1-v^2/c^2}}\left\{ic\frac{dm}{dt}, \mathbf{f}\right\}, \text{ from the above definition of } \mathbf{f}. \quad (28.14)$$

Since $\mathbf{P} = m_0\mathbf{V}$ and m_0 is a constant, then we also have

$$\mathbf{F} = m_0\frac{d\mathbf{V}}{d\tau}, \quad (28.15)$$

which will come in handy shortly.

Now look at the scalar quantity $\mathbf{V} \cdot \mathbf{V}$ (see Chapter 14 for the scalar product of vectors). From equation (28.11),

$$\mathbf{V} \cdot \mathbf{V} = \frac{1}{1-v^2/c^2}\left[(ic)^2 + \mathbf{v} \cdot \mathbf{v}\right],$$

$$= \frac{1}{1-v^2/c^2}\left[-c^2 + v^2\right], \text{ using } i^2 = -1 \text{ and } \mathbf{v} \cdot \mathbf{v} = v^2,$$

$$= -c^2. \quad (28.16)$$

So the scalar product of $\mathbf{V}$ with itself is a constant and consequently differentiating this product must be zero:

$$\mathbf{V} \cdot \frac{d\mathbf{V}}{d\tau} = 0. \quad (28.17)$$

Substituting in from equation (28.15) then gives:

$$\mathbf{V} \cdot \mathbf{F} = 0.$$

At this stage it is now possible to use **V** from equation (28.11) with **F** from equation (28.14):

$$\frac{1}{\sqrt{1-v^2/c^2}}(ic, \mathbf{v}) \cdot \frac{1}{\sqrt{1-v^2/c^2}}\left\{ic\frac{dm}{dt}, \mathbf{f}\right\} = 0,$$

which simplifies to

$$-c^2\frac{dm}{dt} + \mathbf{v} \cdot \mathbf{f} = 0.$$

The last term here, $\mathbf{v} \cdot \mathbf{f}$, is the classical definition of the rate at which the force is doing work, so the work done by the force **f** over the time interval t_1 to t_2 must be:

$$\int_{t_1}^{t_2} c^2 \frac{dm}{dt} dt = c^2 m_2 - c^2 m_1,$$

remembering that mass is not a constant quantity in relativity.

But, from classical physics, the work done is also understood as being the increase in the object's kinetic energy, T, which implies that in relativity theory energy needs to be defined in the form

$$T = mc^2 + \text{constant}.$$

The unknown constant is readily derived by setting the kinetic energy $T = 0$ at $v = 0$, when $m = m_0$ also, in which case:

$$T = (m - m_0) c^2.$$

This is the basis for the equation $E = mc^2$, which states that mass and energy are equivalent: two different forms of the same quantity. The equation can also be turned around, leading to the prediction that an increase in kinetic energy will lead to an increase in the object's mass.

In the classical limit, as $v \ll c$, the familiar $T = \frac{1}{2}m_0 v^2$ is retrieved, since

$$m = m_0 \left(1 - v^2/c^2\right)^{-1/2} \approx m_0 \left(1 + \frac{v^2}{2c^2} + \ldots\right).$$

Relativity theory does not replace the classical physics, it subsumes it.

28.1 Accelerating Spacecraft

The counter-intuitive nature of relativity can be made a bit more concrete by examining the implications of space travel. Hitherto, the analyses have been predicated on the basis of constant-speed relative motion, whereas it is fairly obvious that accelerations are needed if we are get away from the earth, let alone out of the solar system.

To proceed, we need to state the equivalent of the Lorentz transformations for acceleration components. As before, use un-dashed coordinates for the 'static' reference frame and a dashed set for another frame moving to the right with speed v along the x axis. Then the accelerations in the x-direction are related via the following equation [86]:

$$a = a'\left(1 - \frac{v^2}{c^2}\right)^{3/2}.$$

Suppose that the spacecraft containing the moving frame experiences a constant positive acceleration of $a' = g$. This may, for example, be set equivalent to one earth's gravity, thereby being suitable for the well-being of living travellers (uninhabited probes could withstand greater accelerations).

In the static frame, then, the quantity a will be the rate of change of v with respect to time t, so that

$$\frac{dv}{dt} = g\left(1 - \frac{v^2}{c^2}\right)^{3/2}, \qquad (28.18)$$

which has integral

$$gt = \int \frac{dv}{(1 - v^2/c^2)^{3/2}},$$
$$= \frac{v}{\sqrt{1 - v^2/c^2}}, \quad \text{on integration}[9].$$

Rearranging to get an equation for v in terms of t then results in:

$$v = \frac{cgt}{\sqrt{c^2 + g^2 t^2}}, \qquad (28.19)$$

setting $v = 0$ at $t = 0$.

For very large values of time t, it can be seen that $v \to c$ but never reaches it.

By way of example, suppose that $v = 0.95c$ is considered a suitable top speed for the spacecraft; it will then take nearly 3 years (as measured from earth) of continuous acceleration at $g = 9.8$ ms^{-2} to reach this point[10], with another 3 years to decelerate to $v = 0$ at the destination. On the other hand, once at a decent constant cruising speed, Fitzgerald contraction means that the residual distance to the destination will be appreciably foreshortened from the perspective of the moving travellers.

Using $v = dx/dt$ and integrating to obtain x as a function of time[11], ends up with:

$$x - x_0 = \frac{c}{g}\left\{\sqrt{c^2 + g^2 t^2} - c\right\}, \qquad (28.20)$$

---

[9] To integrate, substitute $v = c\sin\eta$, which gives the integral $c \int \sec^2 \eta \, d\eta$. This then equals $c\tan\eta$, since the derivative of $\tan\eta$ is $\sec^2\eta$.
[10] Where the energy is to come from is another matter.
[11] For this integral, use the substitution $gt = c\sinh\theta$.

28.1. ACCELERATING SPACECRAFT

assuming initial conditions such that $x = x_0$ at $t = 0$. Continuing with the above numeric example, the spacecraft will have travelled slightly more than 2 light years distance-wise after the 3 years of acceleration, equivalent to about halfway from earth to its nearest star, Proxima Centauri.

Rearranging equation (28.20), it may be cast in the following form,

$$\left(x - x_0 + \frac{c^2}{g}\right)^2 - c^2 t^2 = \left(\frac{c^2}{g}\right)^2, \tag{28.21}$$

which describes a hyperbolic trajectory [87] for the spacecraft in the 'static' reference frame. We'll come back to the implications of this trajectory later on, but for now just simplify the next bit of algebra by setting $x_0 = 0$.

Equation (28.19) provides the speed versus time from the perspective of a distant observer (on earth, say). But what do the on-board travellers (if any are present) say regarding their own path over time? Given equation (28.19) for v as a function of time t in the static frame, combine with equation (28.10) to get the proper time τ as a function of t:

$$d\tau = \sqrt{1 - \frac{v^2}{c^2}}\, dt,$$

$$= \frac{c\, dt}{\sqrt{c^2 + g^2 t^2}}.$$

Integrating this and setting $\tau = 0$ at $t = 0$, we have

$$t = \frac{c}{g} \sinh\left(\frac{g\tau}{c}\right). \tag{28.22}$$

With $t = 3$ years, we can determine that $\tau = 1.8$ years, so the journey to cruising speed takes a bit less time from the perspective of the travellers than from those on earth.

It is now possible to express x and v as functions of the proper time τ by substituting t from equation (28.22) into equations (28.20) and (28.19) respectively. That is,

$$x = \frac{c^2}{g} \left[\cosh\left(\frac{g\tau}{c}\right) - 1\right], \tag{28.23}$$

$$v = c \tanh\left(\frac{g\tau}{c}\right). \tag{28.24}$$

The latter asymptotes to c as $\tau \to \infty$, consistent with the predictions of equation (28.19).

How far do the accelerating observers think they have travelled? From reference [88], at any time t we can consider an inertial frame that is momentarily at rest with respect to the travellers. In which case, we can use the inverse Lorentz transformation,

$$x = \frac{x' + vt'}{\sqrt{1 - v^2/c^2}},$$

$$= \frac{x'}{\sqrt{1 - v^2/c^2}}, \tag{28.25}$$

setting $t' = 0$ for events that are static in the moving frame.

However, from equation (28.19) for v, it is found that

$$\sqrt{1 - v^2/c^2} = \sqrt{1 - \frac{1}{c^2}\left(\frac{c^2 g^2 t^2}{c^2 + g^2 t^2}\right)} = \frac{1}{\sqrt{1 + g^2 t^2/c^2}}.$$

Therefore, from equation (28.25), we have

$$x' = x\sqrt{1 - v^2/c^2} = \frac{x}{\sqrt{1 + g^2 t^2/c^2}},$$

and equation (28.20) (with $x_0 = 0$) can be used to substitute for x, whereby we finally end up with:

$$x' = \frac{c^2}{g}\left\{1 - \frac{c}{\sqrt{c^2 + g^2 t^2}}\right\}.$$

This can, of course, be expressed in terms of τ instead of t, by means of equation (28.22), which results in

$$x' = \frac{c^2}{g}\left\{1 - \frac{1}{\cosh(g\tau/c)}\right\}.$$

This states that for large values of τ (or t, for that matter), $x' \to c^2/g$ — a fixed asymptotic value. From the point of view of the travellers, then, the sustained acceleration may not appear to have delivered very much in terms of distance actually travelled, and with diminishing gains over time.

An example of the hyperbolic trajectory defined by equation (28.21) is shown as the purple line in Figure 28.1, here using the specific instance of $x_0 = c^2/g$. The units of x along the bottom axis are multiples of c^2/g, while the vertical axis plots ct. Time is assumed to increase from zero.

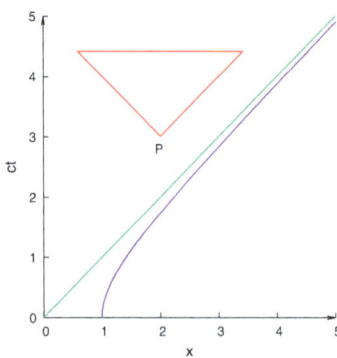

Figure 28.1: Example of the trajectory for an accelerating spaceship

The green line in Figure 28.1 defines the path of a light ray emanating from $x = t = 0$. It can be seen that for large values of time, the spaceship trajectory

28.1. ACCELERATING SPACECRAFT

approaches more and more closely toward this path (which could also have been inferred from equation (28.20)).

It is not obvious from Figure 28.1 (and was certainly not immediately obvious to me), but the green line represents a horizon for the spaceship inhabitants. 'Horizon', in the sense that while the spaceship is accelerating, it cannot receive communications from anything above the green line (call this domain D for simplicity), since no transmission can exceed light speed. To pin this statement down a bit, suppose a radio transmission or similar is sent out in all directions from the point P. The progress of this signal in x-t space is shown by the red triangle[12], here just spanning some interval forward in time (and ignoring the advanced wave going backward in time). Any message sent at less than or equal to light speed from P must lie inside or on this triangle, and so cannot cross the green line. For the travellers to re-enable incoming communications from domain D, just stop accelerating; their trajectory then will be a line oriented more toward the ct axis.

The accelerating travellers are on a trajectory that shuts out a large chunk of spacetime from their purview! (not what one would expect). More drastic sorts of horizon occur in General Relativity.

---

[12] Correctly, a light 'cone', which should encompass three dimensions in space.

Chapter 29

GENERAL RELATIVITY

EINSTEIN'S field equations for general relativity take the form [89],[90]:

$$R_{\mu\nu} - \frac{1}{2}g_{\mu\nu}R + \Lambda g_{\mu\nu} = -\frac{8\pi G}{c^2}T_{\mu\nu}, \qquad (29.1)$$

inclusive of the cosmological constant Λ (capital Greek 'lambda'), which models the accelerating expansion of the universe (now supported by observations). Here, G is the gravitational constant and c the speed of light.

I am not proposing here to go into great detail as to where these equations come from and what all the various terms mean, but we can gain some flavour of what the equation does by means of a more superficial scamper around the margins, so to speak.

The first point to appreciate is that the left-hand-side of the above equation characterises geometry (and for a bit more elaboration on this, see Appendix M), while the right-hand-side of the equation defines the distribution of matter and energy. So matter and energy between them act to alter the geometry — typically creating a curved space-time. The geometry then acts to constrain the paths taken by matter and energy.

The next point is that the description of the geometry is encapsulated in the quantity $g_{\mu\nu}$. This relates to ds^2 by means of the following summation:

$$ds^2 = \sum_{\mu,\nu} g_{\mu\nu} dx^\mu dx^\nu,$$

and ds^2 is, as before, an increment in arc-length, but now in space-time rather than in space alone. The dx^μ and dx^ν define the increments in space and time coordinates and the μ, ν (Greek 'mu' and 'nu') indices run from 1 to 4 (or zero to 3, depending on which convention is adopted).

One can think of this ds^2 equation as a kind of 'super-Pythagoras', and the easiest way of appreciating its form is to write down ds^2 for the particular case of a flat Euclidean space-time from equation (28.8):

$$ds^2 = c^2 dt^2 - \left(dx^2 + dy^2 + dz^2\right), \qquad (29.2)$$

in which the x, y, z stand for the usual space coordinates, t is the time coordinate and c is the speed of light. So $dx^1 = c\,dt$, $dx^2 = dx$, and so on (and where in this case the superscripts do not mean powers but are simply discriminatory integers; admittedly a bit confusing, but the reasons for the convention can be found in [90]).

To obtain $g_{\mu\nu}$, equation (29.1) needs to be solved, given the density of matter and energy $T_{\mu\nu}$ (this is usually a non-trivial exercise). With $g_{\mu\nu}$ in hand, the dynamics of matter within that geometry can be obtained by minimising the integral

$$\int ds = \int \sqrt{\sum g_{\mu\nu}dx^\mu dx^\nu},$$

thus bringing in the Euler-Lagrange equations once again (Section 23).

The paths of light rays within that geometry are obtained from the null geodesics, for which $ds^2 = 0$ (actually solving for the path requires the concept of parallel displacement [90]). This enables the determination of the deflection of light by a gravitational field, for instance in the vicinity of a massive object. Such deflection was used as the first experimental support for general relativity. Returning to the definition of proper time $d\tau = ds/c$ from Chapter 28, the null geodesic $ds = 0$ then implies that a photon's clock (assuming such a thing could exist) would register zero passage of time. So time does not exist for a photon: it can span the entire universe in no time at all [91]. Or, to put it another way, distance does not exist for an electromagnetic wave.

Disturbances in space and time — gravitational waves — also travel on null geodesics [90].

Einstein's field equations for general relativity are currently the only known means for exploring the geometry and evolutionary behaviour of the entire universe.

29.1 Centrifugal Acceleration

If you tie a conker to a length of string and whirl it round your head, you can feel the tension in the string. If the string breaks, or if you let go, the conker will shoot off into the distance. This entirely real force, or acceleration, is a consequence of the rotation and has attracted various explanations at different times. Newton considered the force to be a consequence of rotation relative to absolute space[1]. Mach[2] disagreed; he attributed the force to the effects of all the distant matter in the universe [92]. In neither case was there any agreement as to how the effect was propagated.

---

[1] His well-known bucket experiment established that rotation relative to the bucket was not responsible. In this experiment, he filled a bucket with water, tied a rope to the handle and the other end of the rope to a beam. By twisting up the rope, he was then able to see what happened when the whole lot was allowed to spin freely. Initially, the bucket rotated but the water did not and the surface of the water remained flat. Subsequently, friction between the bucket and the water led to the water rotating as well, which gained a concave surface. Abruptly stopping the bucket left the water still rotating and still with a concave surface.

[2] Ernst Mach, 1838 to 1916.

29.1. CENTRIFUGAL ACCELERATION

What does general relativity say about this? We can find out by creating the relevant metric and determining the dynamic consequences. Start from the flat space metric, equation (28.8):

$$ds^2 = c^2 dt^2 - \left(dx^2 + dy^2 + dz^2\right),$$

and change from Cartesian x-y coordinates to polar coordinates using $x = r\cos\theta$, $y = r\sin\theta$. This gives:

$$ds^2 = c^2 dt^2 - \left(dr^2 + r^2 d\theta^2 + dz^2\right). \tag{29.3}$$

This still describes flat space, just in a different coordinate system. Now define $\theta = \phi + \omega t$, for angular rate ω in the x-y plane. Therefore, $d\theta = d\phi + \omega dt$, thus resulting in a new metric in the rotating coordinates:

$$ds^2 = \left(c^2 - \omega^2 r^2\right) dt^2 - \left(dr^2 + r^2 d\phi^2 + 2\omega r^2 d\phi\, dt + dz^2\right). \tag{29.4}$$

We can imagine the new polar coordinate system r-ϕ as being embedded in a disc rotating about the z-axis. For the subsequent discussion, it helps also to visualise the experiences of an ant or some other inhabitant of that disc; we can expect that if the rotation speed is high enough, the ant will have some difficulty in preventing itself from being spun off altogether.

Before having a look at the dynamics, divert to the concept of *proper time*. As stated in Chapter 28, this is defined in infinitesimal form as

$$d\tau = \frac{ds}{c},$$

and (to reiterate) is the time interval between events that are occurring at the same point in the three-dimensional space to which the observer is attached. It is, if you like, a measurement of 'local time'. The more familiar quantity dt is 'coordinate time', a convenient mathematical time marker (the rotating disc example will hopefully clarify matters).

So in the non-rotating frame, given by the metric in equation (29.3),

$$d\tau_{nr} = dt,$$

setting $dr = d\phi = dz = 0$, consistent with events being at the same spatial location. On the other hand, in the rotating frame defined by the metric in equation (29.4),

$$d\tau_r = \frac{1}{c}\sqrt{c^2 - \omega^2 r^2}\, dt = \sqrt{1 - \frac{\omega^2 r^2}{c^2}}\, dt.$$

Comparing $d\tau_{nr}$ and $d\tau_r$, it can be seen that the latter is less than the former so clocks[3] will run differently in the two different coordinate systems. This statement can be made a bit more specific in terms of measurable events: suppose

---

[3] The word 'clock' in this sense does not just mean a human-made time-keeping device. It means anything that registers the passage of time: atomic vibrations, photon frequencies, life-spans of fundamental particles, and so on.

that the ant on the disc emits light at a specific frequency ν_r (here continuing with the same subscript convention). In some local time interval $\Delta\tau_r$, then, n cycles of that radiation will be emitted. In the non-rotating frame, n cycles will still be received, but in this case $n = \nu_{nr}\Delta\tau_{nr}$ for a different frequency ν_{nr}. Therefore,

$$n = \nu_{nr}\Delta\tau_{nr} = \nu_r\Delta\tau_r,$$

so that

$$\nu_{nr} = \frac{\Delta\tau_r}{\Delta\tau_{nr}}\nu_r = \nu_r\sqrt{1 - \frac{\omega^2 r^2}{c^2}}. \tag{29.5}$$

Since the square-rooted quantity is less than unity, it can be inferred that the received frequency will be *less* than when emitted — it is red-shifted, meaning moved towards the red end of the light spectrum. From the point of view of an external observer away from the rotating disc, it appears as if atoms on the disc are vibrating in slow motion: their local clocks look like they are running more slowly. The same effect is observed if the light is emitted in a strong gravitational field (the sun, for example), and received in a less strong field (the earth); this is termed gravitational red-shift.

In order to determine the dynamics of objects (but not photons) in the rotating coordinate system, we have recourse to the Euler-Lagrange equations resulting from the minimisation of the integral

$$\int ds = \int \left[(c^2 - \omega^2 r^2)\,dt^2 - (dr^2 + r^2 d\phi^2 + 2\omega r^2 d\phi\,dt + dz^2)\right]^{1/2},$$

$$= \int \left[(c^2 - \omega^2 r^2)\,\dot{t}^2 - (\dot{r}^2 + r^2\dot{\phi}^2 + 2\omega r^2\dot{\phi}\dot{t} + \dot{z}^2)\right]^{1/2} ds,$$

$$= \int \sqrt{f\left(\dot{t},\dot{r},\dot{\phi},\dot{z},r\right)}\,ds,$$

the superscript dot here indicating derivative with respect to s and defining

$$f\left(\dot{t},\dot{r},\dot{\phi},\dot{z},r\right) = (c^2 - \omega^2 r^2)\,\dot{t}^2 - \left(\dot{r}^2 + r^2\dot{\phi}^2 + 2\omega r^2\dot{\phi}\dot{t} + \dot{z}^2\right).$$

Now invoke the Euler-Lagrange equations for the four coordinates t, r, ϕ and z:

$$\frac{d}{ds}\left(\frac{\partial f}{\partial \dot{t}}\right) = \frac{\partial f}{\partial t},$$

and similarly for r, ϕ and z. Therefore, treating the coordinates in that order,

$$\frac{d}{ds}\left[\frac{1}{2f}\left\{2\left(c^2 - \omega^2 r^2\right)\dot{t} - 2\omega r^2\dot{\phi}\right\}\right] = 0,$$

$$\frac{d}{ds}\left[\frac{1}{2f}\{-2\dot{r}\}\right] = \frac{1}{2f}\left\{-2\omega^2 r\dot{t}^2 - 2r\dot{\phi}^2 - 4\omega r\dot{\phi}\dot{t}\right\},$$

$$\frac{d}{ds}\left[\frac{1}{2f}\left\{-2r^2\dot{\phi} - 2\omega r^2\dot{t}\right\}\right] = 0,$$

$$\frac{d}{ds}\left[\frac{1}{2f}\{-2\dot{z}\}\right] = 0.$$

29.2. LINEAR ACCELERATION

But since the metric ds^2 is given by equation (29.4), it must be the case that dividing throughout by ds^2 gives $f = 1$, and its derivative with respect to s is zero. This means that the Euler-Lagrange equations simplify to:

$$(c^2 - \omega^2 r^2)\ddot{t} - \omega r^2 \dot{\phi} = A,$$
$$\ddot{r} = \omega^2 r \dot{t}^2 + r\dot{\phi}^2 + 2\omega r \dot{\phi}\dot{t},$$
$$r^2\left(\dot{\phi} + \omega \dot{t}\right) = B,$$
$$\dot{z} = C,$$

introducing constants of integration A, B and C. Combining the first and third of these equations, we get:

$$c^2 \dot{t} = A + \omega B = \text{constant}.$$

If this right-hand-side constant is equated to c, which is equivalent to choosing a particular time scale, we then obtain:

$$\dot{t} = \frac{1}{c},$$

and the differential equation for the radial coordinate r then becomes:

$$\ddot{r} = \frac{\omega^2 r}{c^2} + r\dot{\phi}^2 + \frac{2\omega r \dot{\phi}}{c}.$$

In terms of the proper time $d\tau = ds/c$, this is:

$$\frac{d^2 r}{d\tau^2} = \omega^2 r + r\left(\frac{d\phi}{d\tau}\right)^2 + 2\omega r \frac{d\phi}{d\tau}.$$

Going back to the illustration of the ant on the rotating disc, if the ant itself is not actively moving — so that $d\phi/d\tau = 0$ — it still experiences a radial outward acceleration of $\omega^2 r$. The familiar Coriolis acceleration in the ϕ direction can also be obtained by differentiating the third of the above equations with respect to s, shifting over to proper time and again setting $d\phi/d\tau = 0$:

$$r\frac{d^2\phi}{d\tau^2} + 2\omega \frac{dr}{d\tau} = 0.$$

These accelerations, or forces, experienced by the ant on the rotating disc can thus be regarded as a direct consequence of the purely local distortions of time and space peculiar to that disc. In effect, spinning the disc or whirling the conker round on its string creates its own space-time geometry, which I find to be a quite extraordinary conclusion.

29.2 Linear Acceleration

Earlier on, Section 28.1 touched on the counter-intuitive consequences arising from a linearly-accelerating spaceship, in particular the presence of the horizon, but the analysis stopped short of deriving the corresponding space-time metric ds^2. For the latter, it is possible to start from the fundamental field equations for general relativity, namely $R_{\mu\nu} = 0$, make some assumptions concerning the expected basic form of the metric tensor $g_{\mu\nu}$ and solve for the non-zero components [93]. Rather than follow these lines of thought here, though, it will perhaps be more helpful to work from the following coordinate mappings from [94]:

$$x = \left(x' + \frac{c^2}{g}\right) \cosh\left(\frac{gt'}{c}\right) - \frac{c^2}{g}, \tag{29.6}$$

$$ct = \left(x' + \frac{c^2}{g}\right) \sinh\left(\frac{gt'}{c}\right), \tag{29.7}$$

$$y = y', \quad z = z'.$$

which dovetail very nicely with the approach that can be found in Section 28.1. Here, as usual, x and t are space and time coordinates relative to an external 'static' observer, while the dashed coordinates are measured relative to the accelerating frame.

To retrieve equations (28.23) and (28.22) from (29.6) and (29.7) respectively, set $x' = 0$ — which then concentrates attention entirely on the origin (namely, the spaceship) of the accelerating frame — and replace t' with τ, consistent with the use of proper time for a clock attached to that frame.

Work from the flat-space metric for invariant ds^2, namely:

$$ds^2 = c^2 dt^2 - \left(dx^2 + dy^2 + dz^2\right).$$

From equation (29.7), we find that

$$c\, dt = dx' \sinh\left(\frac{gt'}{c}\right) + dt'\frac{g}{c}\left(x' + \frac{c^2}{g}\right) \cosh\left(\frac{gt'}{c}\right),$$

whereas from equation (29.6),

$$dx = dx' \cosh\left(\frac{gt'}{c}\right) + dt'\frac{g}{c}\left(x' + \frac{c^2}{g}\right) \sinh\left(\frac{gt'}{c}\right).$$

Substituting these into the above equation for ds^2, collecting terms and simplifying eventually gives:

$$ds^2 = c^2 dt'^2 \left(1 + \frac{gx'}{c^2}\right)^2 - \left(dx'^2 + dy'^2 + dz'^2\right), \tag{29.8}$$

valid for $|x'| < c^2/g$.

This metric is sometimes used in the context of weak gravitational fields. To see the correspondence, write the linear acceleration g that we have been using hitherto in this section in the form

$$g \approx \frac{GM}{r^2},$$

(if this is unfamiliar, see Section 11.4) change its sign so that the 'force' is attractive rather than repulsive, and replace x' with a radial coordinate r. Then,

$$1 + \frac{gx'}{c^2} \approx 1 - \frac{GM}{c^2 r}.$$

Apart from a factor of 2, this is similar to the leading term in the Schwarzschild metric presented in the next section, which characterises the geometry near to a spherically-symmetric mass.

29.3 Black Holes

The final collapse of a sufficiently large star, once all the nuclear fuel is used up, is predicted to produce a black hole. This is a region of space-time with such an intense gravitational field that not even light can escape [2]. It is now accepted that most galaxies have a massive black hole at their centre, strongly affecting the motions of nearby stars. The existence of black holes could have been inferred in 1916, when the Schwarzschild[4] metric was published – the first exact solution of the field equations. This metric characterises the geometry in the vicinity of a massive spherical object (such as a star):

$$ds^2 = \left(1 - \frac{2\alpha}{r}\right) c^2 dt^2 - \frac{1}{(1 - 2\alpha/r)} dr^2 - r^2 \left(d\theta^2 + \sin^2\theta \, d\phi^2\right). \quad (29.9)$$

The radial coordinate r is measured from the centre of the massive object, t is time as before and the angle coordinates are θ and ϕ. The constant α has units of length, and involves Newton's gravitational constant G, the mass M of the star and the speed of light c [90]; specifically:

$$\alpha = \frac{GM}{c^2}.$$

For large r, well away from the star, this metric simplifies to the flat-space solution, but look at what happens if r is less than 2α: the sign of $1 - 2\alpha/r$ becomes negative, so the time and radial coordinates in ds^2 both swap sign. Normally, this critical radius 2α is inside the star's mass, but for a black hole the boundary — its event horizon — is in space and so accessible to a space traveller. From the change in coordinate sign it can be inferred that once inside the critical radius, one can move freely in time[5] but in only one direction in r.

A more detailed examination of the history of an object (a spaceship, say) moving radially inwards toward the event horizon $r = 2\alpha$ may be found in [90] or [95]. This situation is worth examining in a bit more detail, since the narratives of what happens differ significantly depending on whether the observer is watching from a distance or is travelling with the spaceship.

The key differential equations for such a purely radial motion can be obtained from the metric in equation (29.9), in conjunction with the usual Euler-Lagrange

---
[4]Karl Schwarzschild, 1873 to 1916.
[5]Whether this is actually so is at present undetermined.

equations, resulting in [90]:

$$\frac{dt}{ds} = \frac{1}{c}\left(1 - \frac{2\alpha}{r}\right)^{-1},$$

$$\left(\frac{dr}{ds}\right)^2 = \frac{2\alpha}{r},$$

in terms of the path length s relative to some suitable starting point.

Since it is expected that r should diminish with s, the negative square root is appropriate for the second of these equations, in which case we have

$$\frac{dr}{ds} = -\sqrt{\frac{2\alpha}{r}}.$$

Look first at what happens to r in terms of the proper time τ, where $d\tau = ds/c$ as usual. Therefore,

$$\frac{dr}{d\tau} = -c\sqrt{\frac{2\alpha}{r}}.$$

Integrating this is straightforward enough, resulting in

$$r^{3/2} = r_0^{3/2} + \frac{3}{2}c\sqrt{2\alpha}\,(\tau_0 - \tau),$$

in terms of initial values r_0, τ_0. There is nothing in this equation to indicate that r cannot decay smoothly to zero as the proper time increases. In fact, if we term τ_f the proper time at which r does reach zero, then

$$r^{3/2} = \frac{3}{2}c\sqrt{2\alpha}\,(\tau_f - \tau).$$

From this, we can infer that any traveller in that spaceship heading radially into the black hole will experience nothing particularly untoward when passing through the event horizon (apart from potentially destructive tidal forces).

From the perspective of a distant observer, however, things look very different; to see this, we need to solve for r as a function of time t:

$$\frac{dr}{dt} = \frac{dr/ds}{dt/ds} = -c\sqrt{\frac{2\alpha}{r}}\left(1 - \frac{2\alpha}{r}\right).$$

Turning this upside down to get t as a function of r gives

$$c\frac{dt}{dr} = -\frac{1}{\sqrt{2\alpha}}\left(\frac{r^{3/2}}{r - 2\alpha}\right).$$

This can be integrated exactly [36], but it is then difficult to discern the interesting behaviour from among the resulting welter of square roots. Since the domain of interest is when r approaches 2α, it makes sense to examine this

29.3. BLACK HOLES

particular region in more detail. Thus, set $r = 2\alpha + \rho$ for a new variable ρ, and consider ρ as a small positive quantity. That is,

$$c\frac{dt}{d\rho} = -\frac{2\alpha}{\rho}\left(1 + \frac{\rho}{2\alpha}\right)^{3/2},$$

$$= -\frac{2\alpha}{\rho}\left[1 + \frac{3\rho}{4\alpha} + \frac{3\rho^2}{32\alpha^2} + \ldots\right],$$

expanding in a Taylor series for small $\rho/2\alpha$.

Integrating this is straightforward enough:

$$c(t - t_0) = -2\alpha\left[\log_e \rho + \frac{3\rho}{4\alpha} + \frac{3\rho^2}{64\alpha^2} + \ldots\right]_{\rho_0}^{\rho}.$$

Therefore, provided both ρ and ρ_0 can be considered as small quantities, we end up with the approximation,

$$\rho \approx \rho_0 e^{-c(t-t_0)/2\alpha}.$$

For someone watching at a distance from the black hole, the spaceship is never perceived to pass through the horizon, instead ending up perpetually frozen on its boundary.

Numerical integration of the differential equations for ct, r and τ with respect to s supports these conclusions. As an example, using $2\alpha = 1$ results in Figure 29.1, here showing r as a function of s and r as a function of ct.

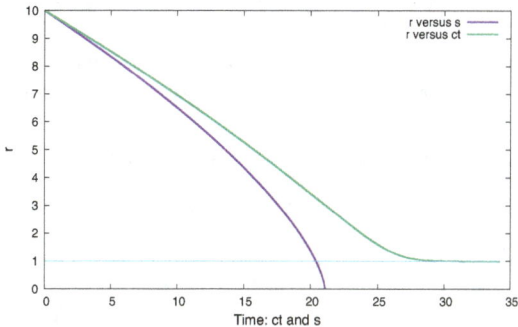

Figure 29.1: Behaviour of r with respect to s and ct

From the viewpoint of a distant observer, then, the time to reach the horizon becomes infinite — the horizon is never reached. On the other hand, for a local observer travelling with that object, the horizon is passed at a finite local ('proper') time. Also, by analogy with equation (29.5), it can be inferred that the distant observer will see any electromagnetic emissions from the spaceship as being shifted toward the red end of the spectrum, thus becoming both redder and fading out in energy terms (since $\mathcal{E} = h\nu$), as $r \to 2\alpha$. That is, if ν_s and

ν_d stand for the frequencies seen at the spaceship and by the distant observer respectively, then

$$\nu_d = \nu_s \sqrt{1 - \frac{2\alpha}{r}}.$$

This gravitational red shift effect is picked up in a related context in Section 29.4.

Returning to the metric in equation (29.9), the apparent singularity in the dr^2 term at $r = 2\alpha$ is due to the choice of coordinates — both r and t need to be replaced in order to obtain a more uniformly valid metric.

29.4 GPS Timing Corrections

It was mentioned in Section 1 that accurate receiver localisation using GPS requires two relativistic corrections [96], and it is instructive to see what these amount to in approximate numerical terms, especially given the expected positional accuracy on the ground.

There are two relativistic effects involved, both from general relativity. The first of these is a consequence of clocks running *slower* on the satellite relative to the receiver on the ground, due to the greater orbital speed of the satellite (relative to the ground). This correction is given by

$$d\tau_s = \sqrt{1 - \frac{v^2}{c^2}}\, dt,$$

where v is the satellite speed, dt stands for coordinate time and $d\tau_s$ for proper time. The subscript $_s$ here stands for 'speed'.

This expression for the proper time can be obtained using the methods of Section 29.1, with $v = \omega r$ for a circular orbit (an elliptical orbit can simply be accommodated by allowing for a variable v).

Since the satellite speed is much less than c, $d\tau_s$ may be approximated by:

$$d\tau_s \approx \left(1 - \frac{v^2}{2c^2}\right) dt.$$

The other timing correction stems from the difference in gravitational potentials, in that the earth's gravity is weaker for the satellite than for the receiver nearer to the earth surface. So clocks run *faster* on the satellite relative to the receiver. The key equation here may be taken from the Schwarzschild metric[6] in Section 29.3, with the proper time given by

$$d\tau = \sqrt{1 - \frac{2GM}{c^2 r}}\, dt, \tag{29.10}$$

---

[6] I have not yet been able to determine whether the Kerr metric [90] for a *rotating* earth is more appropriate, or if the difference from the Schwarzschild metric is too small to worry about in this context.

29.4. GPS TIMING CORRECTIONS

where r is the distance from the centre of the earth. This may also be approximated:

$$d\tau \approx \left(1 - \frac{GM}{c^2 r}\right) dt.$$

This is the approximate expression for the proper time for an object at distance r from the earth's centre. What we need is the difference in proper times between a satellite at distance r_s and the receiver at distance r_e. That is,

$$d\tau_g \approx \left\{1 - \left(\frac{GM}{c^2 r_s} - \frac{GM}{c^2 r_e}\right)\right\} dt = \left\{1 + \frac{GM}{c^2}\left(\frac{1}{r_e} - \frac{1}{r_s}\right)\right\} dt,$$

which is the same as working from the difference in gravitational potentials.

The timing *differences*, proper time minus coordinate time, is then given by the following equations[7]:

$$\text{Effect of orbital speed: } d\tau_s - dt = -\frac{1}{2}\frac{v^2}{c^2} dt,$$

$$\text{Effect of satellite altitude: } d\tau_g - dt = \frac{GM}{c^2}\left(\frac{1}{r_e} - \frac{1}{r_s}\right) dt.$$

The first of these is negative: clocks run slower; while the second is positive: clocks run faster. To put some rough and ready numbers into these, use $GM = 3.986 \times 10^{14}$, $c = 3 \times 10^8$, $r_e = 6378137$, $r_s = 2.6378 \times 10^7$ and $v = 4000$ (all in SI units and the last two figures approximately appropriate for a GPS satellite[8]). We then find that

$$d\tau_s - dt \approx -8.88 \times 10^{-11} dt, \text{ or about -7.7 microseconds per day,}$$

$$d\tau_g - dt \approx 5.26 \times 10^{-10} dt, \text{ or about 45.48 microseconds per day.}$$

The net effect of ignoring these relativistic timing corrections would be to incur a timing error of about 38.1 microseconds[9] in one day [97], [98]. But although such an error seems tiny, it has a significant effect on the receiver localisation error. To see this, we turn to the basic time-difference-of-arrival equation that is used to determine the receiver's location [99]:

$$(x - \xi_i)^2 + (y - \eta_i)^2 + (z - \zeta_i)^2 = c^2 (\Delta t_i - d)^2, \tag{29.11}$$

where x, y, z are the Cartesian coordinates of the receiver (the 'unknowns'); ξ_i, η_i, ζ_i are the coordinates of satellite i (known); Δt_i is the time difference at the receiver from satellite i (which can be worked out in the receiver); and d is a timing correction (also unknown, and needed because the receiver's clock is generally much less accurate than those on the GPS satellites).

---

[7] These may also be obtained from the Schwarzschild metric, equation (29.9), but generalised for an orbiting object using $\theta = \theta_0 + \omega t$, as in Section 29.1.
[8] In fact, both r_s and v vary for an elliptical orbit.
[9] A microsecond is one millionth of a second.

Since there are four unknown quantities, it can be appreciated that at least four satellites need to be visible to the receiver for a solution to be determinable[10].

There is no need here to discuss how equation (29.11) can be solved; the important point is that any timing error will be multiplied by c when determining location. So the 38.1 microseconds per day will be roughly equivalent to some 11.4 kilometres in spatial offset terms — rather larger than we are accustomed to with GPS. To obtain an expected spatial accuracy of 15 metres or so, the timing errors need to be no more than about 50 nanoseconds[11] per day.

In the interests of simplicity, the above discussion has deliberately omitted several corrective refinements that are part of the real GPS design. One of these refinements is a more accurate expression for the earth's gravitational potential; this appears in equation (29.10) as the GM/r term, but should include at least the J_2 primary oblateness harmonic (reflecting the fact that the earth is not actually spherical). Other refinements include corrections for a receiver being on (or near) a rotating rather than static earth. A more complete discussion can be found in [100].

Without knowledge of the equations for both special and general relativity, GPS would be unable to provide the expected accurate localisation.

---

[10] And your mobile phone or GPS receiver then needs to solve equation (29.11) in order to work out where it is.

[11] A nanosecond is one billionth of a second, with a billion equal to one thousand million.

Chapter 30

QUANTUM MECHANICS

IT was a bit unfair to throw Schrödinger's[1] equation into the start of Chapter 17 without any sort of explanation[2], since elsewhere I've (mostly) attempted to provide some understanding as to where equations come from. So this chapter goes some way toward filling the gap, especially since this equation has proved so immensely successful in describing the behaviour of matter at atomic scales.

There seems to be no general agreement as to how Schrödinger originally came up with the equation that is named after him, and nor is there any consensus in terms of a derivation from classical physics (meaning pre-relativistic physics). But it is possible to show the plausibility of the equation, given links back to electromagnetism and Einstein's special relativity [101], [102].

We start from the wave equation 25.8, using one spatial dimension for simplicity:

$$\frac{\partial^2 E}{\partial x^2} = \frac{1}{c^2}\frac{\partial^2 E}{\partial t^2}, \qquad (30.1)$$

which describes the behaviour of an electro-magnetic wave in a vacuum, c being the speed of light. This will have plane-wave solutions in the complex domain of the form

$$E(x,t) = E_0 e^{i(kx-\omega t)},$$

(see equation (25.12)), where the wave number k relates to the wavelength λ by $k = 2\pi/\lambda$, and $\omega = 2\pi\nu$, ν being the frequency. Substituting this into equation (30.1) then results in the well-known relation linking frequency and wavelength for electromagnetic radiation:

$$k = \frac{\omega}{c}, \text{ or } c = \nu\lambda.$$

---

[1] Erwin Schrödinger, 1887 to 1961.
[2] I've never liked taking equations or assumptions entirely on trust, although sometimes it is necessary in order to make progress.

Now bring in Einstein's equation linking energy and frequency for a photon (a 'particle', or quantum, of light or radiation), namely $\mathcal{E} = h\nu$, where h is Planck's constant; and also the Compton[3] relation between momentum p and wavelength λ: $p = h/\lambda$. Therefore,

$$k = \frac{2\pi}{\lambda} = \frac{2\pi p}{h}, \text{ and } \omega = 2\pi\nu = \frac{2\pi\mathcal{E}}{h}.$$

So rewriting $E(x,t)$:

$$E(x,t) = E_0 e^{i(px - \mathcal{E}t)/\hbar},$$

where $\hbar = h/2\pi$ as the usual abbreviation.

If this is now substituted into equation (30.1), we find that

$$\left(\frac{\partial^2}{\partial x^2} - \frac{1}{c^2}\frac{\partial^2}{\partial t^2}\right) E_0 e^{i(px - \mathcal{E}t)/\hbar} = 0,$$

or $\quad -\frac{1}{\hbar^2}\left(p^2 - \frac{\mathcal{E}^2}{c^2}\right) E_0 e^{i(px - \mathcal{E}t)/\hbar} = 0,$ \hfill (30.2)

which implies that

$$\mathcal{E}^2 = p^2 c^2. \tag{30.3}$$

This may be recognised as a special case of the relativistic total energy, namely,

$$\mathcal{E}^2 = p^2 c^2 + m^2 c^4, \tag{30.4}$$

for a particle with zero rest mass, in which $m = 0$.

We now assume that frequency and energy, wavelength and momentum are related in exactly the same way for particles as for photons and so try to derive a wave equation for non-zero rest mass particles. Which means that we need to end up with a dynamic equation that results in equation (30.4) rather than (30.3). In other words, we would like to get something of the form

$$-\frac{1}{\hbar^2}\left(p^2 - \frac{\mathcal{E}^2}{c^2} + m^2 c^2\right) \Psi_0 e^{i(px - \mathcal{E}t)/\hbar} = 0,$$

in place of equation (30.2), here shifting over to a wave function Ψ in place of E.

But this can be obtained from the differential equation

$$\left(\frac{\partial^2}{\partial x^2} - \frac{1}{c^2}\frac{\partial^2}{\partial t^2} - \frac{m^2 c^2}{\hbar^2}\right) \Psi_0 e^{i(px - \mathcal{E}t)/\hbar} = 0,$$

or

$$\frac{\partial^2 \Psi}{\partial x^2} - \frac{m^2 c^2}{\hbar^2}\Psi = \frac{1}{c^2}\frac{\partial^2 \Psi}{\partial t^2}, \tag{30.5}$$

---
[3] Arthur Compton, 1892 to 1962.

here defining
$$\Psi = \Psi_0 e^{i(px - \mathcal{E}t)/\hbar}.$$

The Klein-Gordon[4] equation (30.5), though, is relativistic, whereas we need a non-relativistic equivalent since the Schrödinger equation is non-relativistic. So approximate the total energy for small values of p relative to mc, consistent with speeds being small relative to that of light:

$$\mathcal{E} = \sqrt{p^2 c^2 + m^2 c^4} = mc^2 \sqrt{1 + \frac{p^2}{m^2 c^2}}$$

$$\approx mc^2 \left(1 + \frac{p^2}{2m^2 c^2} + \ldots\right) \approx mc^2 + \frac{p^2}{2m} + \ldots.$$

The last term here may be recognised as the classical kinetic energy $T = \frac{1}{2}mv^2$, using $p = mv$ in terms of velocity v.

Now use this approximation in the above definition of Ψ:

$$\Psi \approx \Psi_0 e^{i(px - mc^2 - Tt + \ldots)/\hbar},$$
$$\approx e^{-imc^2 t/\hbar} \psi(x, t),$$

defining $\psi(x,t) = \Psi_0 e^{i(px - Tt)/\hbar}$ and ignoring the higher-order terms.

Differentiate Ψ with respect to time:

$$\frac{\partial \Psi}{\partial t} = -\frac{i}{\hbar} mc^2 e^{-imc^2 t/\hbar} \psi + e^{-imc^2 t/\hbar} \frac{\partial \psi}{\partial t},$$

$$\frac{\partial^2 \Psi}{\partial t^2} = \left[-\frac{m^2 c^4}{\hbar^2} e^{-imc^2 t/\hbar} \psi - \frac{2i}{\hbar} mc^2 e^{-imc^2 t/\hbar} \frac{\partial \psi}{\partial t}\right] + e^{-imc^2 t/\hbar} \frac{\partial^2 \psi}{\partial t^2},$$

$$\approx \left[-\frac{m^2 c^4}{\hbar^2} e^{-imc^2 t/\hbar} \psi - \frac{2i}{\hbar} mc^2 e^{-imc^2 t/\hbar} \frac{\partial \psi}{\partial t}\right],$$

dropping the small last term, since c is large.

Substitute this approximate second-time-derivative term into equation (30.5):

$$\left(\frac{\partial^2}{\partial x^2} - \frac{m^2 c^2}{\hbar^2}\right) e^{-imc^2 t/\hbar} \psi \approx \frac{1}{c^2} \left[-\frac{m^2 c^4}{\hbar^2} e^{-imc^2 t/\hbar} \psi - \frac{2i}{\hbar} mc^2 e^{-imc^2 t/\hbar} \frac{\partial \psi}{\partial t}\right]$$

$$\approx -\frac{m^2 c^2}{\hbar^2} e^{-imc^2 t/\hbar} \psi - \frac{2im}{\hbar} e^{-imc^2 t/\hbar} \frac{\partial \psi}{\partial t}.$$

The terms linear in ψ cancel out, and the common factor $e^{-imc^2 t/\hbar}$ may also be removed since this multiples every term present. This leaves:

$$\frac{\partial^2 \psi}{\partial x^2} = -\frac{2im}{\hbar} \frac{\partial \psi}{\partial t},$$

which may be rearranged to give a more familiar form:

$$i\hbar \frac{\partial \psi}{\partial t} = -\frac{\hbar^2}{2m} \frac{\partial^2 \psi}{\partial x^2},$$

---

[4]Oskar Klein, 1894 to 1977 and Walter Gordon, 1893 to 1940.

and this is a special case of the more general Schrödinger equation inclusive of a potential energy U [103]:

$$i\hbar \frac{\partial \psi}{\partial t} = -\frac{\hbar^2}{2m}\nabla^2 \psi + U\psi,$$

where in Cartesian x-y-z coordinates the ∇^2 quantity stands for

$$\frac{\partial^2}{\partial x^2} + \frac{\partial^2}{\partial y^2} + \frac{\partial^2}{\partial z^2},$$

an operator in three spatial dimensions.

Finally, due to its similarity to the classical equivalent, the Hamiltonian operator H is defined as:

$$H = -\frac{\hbar^2}{2m}\nabla^2 + U, \qquad (30.6)$$

thus resulting in the Schrödinger equation in its usual form:

$$i\hbar \frac{\partial \psi}{\partial t} = H\psi. \qquad (30.7)$$

The wave function ψ is a complex quantity, and the probability of finding a particle at a specific place and time is given by $\psi\psi^*$, ψ^* being the complex conjugate. This probabilistic aspect of quantum mechanics is quite different from the more familiar classical world, in which some object is either here or it isn't — and not possibly here, there or anywhere else. Other than in Section 30.1, I do not intend to delve any further into the fascinating but decidedly weird and counter-intuitive quantum environment, but there is no doubt that the strange behaviours reflect reality as it really is [104], if only because detailed experiments confirm the predictions of the quantum equations to astonishing levels of accuracy.

Returning briefly to the relativistic domain, equation (30.5) (also known as the relativistic Schrödinger equation) describes the behaviour of particles with integer spin (such as π-mesons), as well as photons (with zero rest mass), whereas the Dirac equation

$$i\hbar \frac{\partial \psi}{\partial t} = i\alpha\hbar c \nabla \psi + \beta mc^2 \psi$$

is needed for particles of half-integer spin such as electrons. The quantities α and β both involve 4×4 matrices, while the ∇ vector operator in Cartesian coordinates is

$$\nabla = \left(\frac{\partial}{\partial x}, \frac{\partial}{\partial y}, \frac{\partial}{\partial z}\right).$$

As a final remark here, the Klein-Gordon equation (30.5) admits both advanced- and retarded-wave type solutions (as for the wave equation in Section 25.2), since we can write

$$\Psi \sim e^{i(\omega t - \nu x)},$$

provided that $\omega^2 = c^2(\nu^2 + m^2c^2/\hbar^2)$. On the other hand, the Schrödinger equation (30.7) has only one solution, usually interpreted as propagating forward in time.

However, returning to the calculation of probabilities from $\psi\psi^*$, we find that the equivalent of the Schrödinger equation for the complex conjugate ψ^* is

$$-i\hbar\frac{\partial \psi^*}{\partial t} = H\psi^*,$$

which is just a time-reversed form of equation (30.7) [91].

In effect, calculating the probabilities from $\psi\psi^*$ retrieves both advanced and retarded solutions, albeit in a disguised form!

30.1 Quantum Tunnelling

One of the simpler exact solutions to equation (30.7) serves to illustrate the unusual nature of reality at small scales, and is to do with the ability of particles to tunnel through a barrier that would be completely impassable in classical physics. You put your box of cereal in the cupboard last thing at night, and do not expect to find it in the middle of the kitchen floor at breakfast time — and yet that is exactly what can happen in the quantum domain. Locations of objects cease to be definite, but instead dissolve into a kind of fuzziness, meaning that a particle can be spread out in space; perhaps with a greater probability of being here rather than there — at least until a measurement is made, when things become more definite.

In this sense, Schrödinger's equation is only half of the story: it governs the behaviour of quantum particles when left to themselves, but the moment a measurement is made — such as looking in the cupboard — the probabilities all collapse down into a single value: the particle is *here* with unit probability. Various explanations have been put forward for this 'collapse of the wave function' phenomenon, but understanding what actually goes on remains a mystery. To quote Richard Feynman[5], "... I think I can safely say that nobody understands quantum mechanics" [105], a statement that remains just as valid today.

Let us suppose that we have managed to pin down the location of a particle to somewhere in the region $x < 0$, to the left of the barrier illustrated in Figure 30.1, with a possible position marked schematically by the blue dot. The barrier is rendered simplistically as the cross-hatched 'wall' in this one-dimensional example, and could — for example — stand for an electromagnetic repulsive force keeping the charged blue dot to the left of $x = 0$.

---

[5]Richard Feynman, 1918 to 1988.

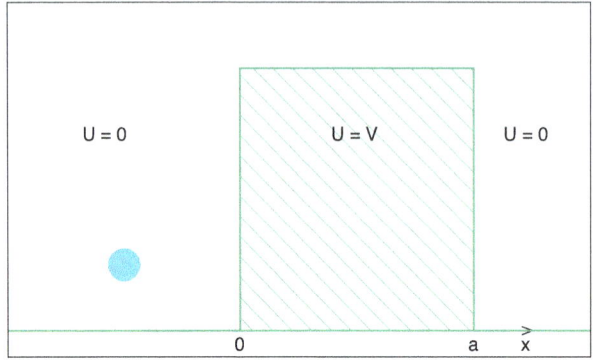

Figure 30.1: Schematic barrier

From equations (30.7) and (30.6), we get the governing wave function ψ by solving:

$$i\hbar \frac{\partial \psi}{\partial t} = -\frac{\hbar^2}{2m} \frac{\partial^2 \psi}{\partial x^2} + U\psi, \tag{30.8}$$

after restricting to the one spatial dimension x, and here the potential U is a function of x only. To solve this, adopt the separability approach from Section 25.2, in which ψ as a function of both t and x is regarded as a product of independent functions of the form:

$$\psi(t, x) = T(t)\phi(x).$$

Substitute this into equation (30.8):

$$i\hbar \frac{dT}{dt} \phi = -\frac{\hbar^2}{2m} T \frac{d^2\phi}{dx^2} + UT\phi,$$

and divide both sides by $T\phi$:

$$i\hbar \frac{1}{T} \frac{dT}{dt} = -\frac{\hbar^2}{2m} \frac{1}{\phi} \frac{d^2\phi}{dx^2} + U = \text{constant}.$$

As mentioned in Section 25.2), the independence of the functions of t and x is preserved by making both sides equal to a constant, as shown. This constant is interpreted as the particle's energy E, so we get two equations linked by E:

$$i\hbar \frac{1}{T} \frac{dT}{dt} = E,$$

$$\frac{\hbar^2}{2m} \frac{1}{\phi} \frac{d^2\phi}{dx^2} - U + E = 0.$$

The first of these is readily solved to give:

$$T(t) = T_0 e^{-iEt/\hbar},$$

30.1. QUANTUM TUNNELLING

for some constant T_0.

For the second differential equation, refer to Figure 30.1. For $x < 0$ and $x > a$, the potential U is zero, whereas for $0 \leq x \leq a$, U takes a constant positive value V. The solution for ϕ therefore needs to be split into three parts:

- $x < 0$:

With $U = 0$, we have the simplified equation

$$\frac{d^2\phi}{dx^2} + \frac{2mE}{\hbar^2}\phi = 0,$$

which has sinusoidal solutions of the form

$$\phi(x) = A\cos wx + B\sin wx,$$

setting $w = \sqrt{2mE}/\hbar$.

- $0 \leq x \leq a$:

This is the more interesting zone inside the barrier, and we assume that $V > E$ so that ordinarily the particle would not have enough energy to surmount the obstacle. The differential equation then is:

$$\frac{d^2\phi}{dx^2} - \frac{2m(V-E)}{\hbar^2}\phi = 0.$$

Write $\mu^2 = 2m(V-E)/\hbar^2 > 0$. The solutions must then be of the form:

$$\phi(x) = Ce^{\mu x} + De^{-\mu x}.$$

The $e^{\mu x}$ solution implies an exponentially increasing probability of finding the particle the further one penetrates into the barrier, which does not seem sensible on physical grounds. So the solution inside the barrier is, simply,

$$\phi(x) = De^{-\mu x},$$

for some constant D.

- $x > a$:

The solution here is of the same character as for $x < 0$, and will be

$$\phi(x) = P\cos wx + Q\sin wx,$$

for some other constants P and Q.

There seem to be a lot of undetermined constants involved here, but the number of independent ones can be reduced by matching ϕ and $d\phi/dx$ across the two boundaries. In any case, it is what goes on inside the barrier that is of most interest, and the wave function here will have the form:

$$\psi(t,x) = De^{-iEt/\hbar}e^{-\mu x},$$

absorbing T_0 into D (both are constants, after all). The probability of finding the particle inside the barrier then must be

$$\text{probability} = \psi\psi^*, \text{ (a superscript } {}^* \Rightarrow \text{ complex conjugate, for which } i \to -i),$$
$$= \left[De^{-iEt/\hbar}e^{-\mu x}\right]\left[De^{iEt/\hbar}e^{-\mu x}\right],$$
$$= D^2 e^{-2\mu x}.$$

So the resulting probability is independent of time. The probability is also not zero, although it will decrease as x increases. This means that there is a finite probability that the particle can reach $x = a$ and so tunnel completely through the barrier. Such predictions have a practical impact, since the above model provides a decent understanding as to what happens in radioactive decay. For example, based on the above results, it is expected that the probability of an α-particle[6] tunnelling out of an atomic nucleus will be given by [102]:

$$p \propto \exp\left(-2\int_{r_1}^{r_2} \frac{\sqrt{2m(V-E)}}{\hbar}dr\right),$$

which agrees with experiments[7].

It may be asked, why do we not see such tunnelling occurring in everyday life? The answer is to do with the magnitude of Planck's constant, since $\hbar \sim 1.054 \times 10^{-34}$ in SI units[8]. At our sort of human length scales, the ensuing probabilities are so small as to be effectively zero.

So the next time your cornflakes go walkies, you'll know what happened.

---

[6]The name given to a composite consisting of two protons and two neutrons.
[7]The symbol $\propto$ means 'proportional to'.
[8]The curly symbol $\sim$ stands for 'of the order of'.

Chapter 31

OPTIMAL CONTROL

IMAGINE the following situation: you are one of a team of engineers at NASA with the task of landing astronauts on the moon, and ensuring their safe return. In contrast to Mars (for example), the moon has no atmosphere, so atmospheric braking of the landing craft using a parachute or similar is not an option. This means that rocket thrusters are essential if the craft is to land in one piece. Everything on that landing craft, including the rocket fuel, must be launched from earth and the more mass that needs to go up, the more powerful the main rocket must be (and the greater the cost involved). A further consideration is that when fuel is used to brake the landing craft, sufficient must remain to enable the astronauts to get back up to the 'mother ship'.

There is, then, a requirement to minimise the amount of thruster fuel required — most especially during the descent to the moon's surface. If not enough is left, the unfortunate astronauts may be stranded.

The human reaction to the situation would probably be similar to braking a car to a stop within some required distance: apply the brakes gradually at first, then increasing or decreasing the pressure as one progresses, using continual visual judgment and experience to achieve the desired outcome. For the moon landing case, this strategy is not optimal — more fuel will be used than is necessary. The actual solution, counter-intuitive though it may be, is to allow the craft to accelerate downward to a certain height and then turn the retro-rockets on at full power. Timing is everything, of course, which is where the mathematics comes in. Before progressing to the relevant equations, first set out the method of Pontryagin [1].

31.1 Pontryagin Formalism

The method of Pontryagin is to use the calculus of variations technique (Chapter 23), with a set of time-dependent Lagrange multipliers λ_j, to minimise the cost integral J for an arbitrary control u, and then choose those controls that will achieve the extremum compatible with the constraints [69]. Suppose that

---

[1] Lev Pontryagin, 1908 to 1988.

the cost integral is of the form

$$J = \int_{t=0}^{T} F(x_i, u, t)\,dt,$$

where t stands for time $\in [0, T]$, x_i denotes the position components, $i \in [1, n]$, and $u(t)$ is the control (here assumed singular for simplicity).

Assume also that the dynamics of the situation is encapsulated in a set of first-order ordinary differential equations of the form:

$$\dot{x}_i = f_i(x_j, u, t); \quad \text{(a function of one or more coordinates, plus } u \text{ and } t\text{)}.$$

The superscript dot denotes time derivative. Then follow the calculus of variations approach with time-dependent Lagrange multipliers to extend the cost integral to become

$$J = \int_{t=0}^{T} \left[F(x_i, u, t) + \sum_i \lambda_i(t)\left(\dot{x}_i - f_i(x_j, u, t)\right) \right] dt,$$

at the same time defining the Lagrangian as

$$\mathcal{L} = F(x_i, u, t) + \sum_i \lambda_i(t)\left(\dot{x}_i - f_i(x_j, u, t)\right). \tag{31.1}$$

Now formulate the Euler-Lagrange equations for the x_i, resulting in the usual set:

$$\frac{d}{dt}\left(\frac{\partial \mathcal{L}}{\partial \dot{x}_j}\right) = \frac{\partial \mathcal{L}}{\partial x_j}, \quad \text{for } j = 1, \ldots, n.$$

which give rise to

$$\dot{\lambda}_j = \frac{\partial F}{\partial x_j} - \sum_i \lambda_i \frac{\partial f_i}{\partial x_j}, \quad \text{for } j = 1, \ldots, n. \tag{31.2}$$

Functional dependencies have been omitted in the interests of compactness. In addition, the $\lambda_i(t)$ are treated as if they are coordinates in their own right, so that

$$\frac{d}{dt}\left(\frac{\partial \mathcal{L}}{\partial \dot{\lambda}_i}\right) = \frac{\partial \mathcal{L}}{\partial \lambda_i}, \quad \text{for } i = 1, \ldots, n,$$

which — from the above form of $\mathcal{L}$ — simply retrieve the original kinematic equations, namely

$$\dot{x}_i = f_i.$$

The penultimate piece is to define the *pseudo-Hamiltonian*, H', such that:

$$H' = F - \sum_i \lambda_i f_i, \tag{31.3}$$

31.1. PONTRYAGIN FORMALISM

and then choose the control u that gives an extremal of H', via

$$\frac{\partial H'}{\partial u} = 0. \tag{31.4}$$

This H' is not as arbitrary as it may appear; it is, in fact, the negative of the Hamiltonian from Chapter 24, which is defined as

$$H = \sum_i \dot{x}_i \frac{\partial \mathcal{L}}{\partial \dot{x}_i} - \mathcal{L},$$

here using x_i in place of q_i. To see this, carry out the partial derivative of $\mathcal{L}$ with respect to one of the coordinate derivatives — $\dot{x}_k$, say. Then, from equation (31.1),

$$\frac{\partial \mathcal{L}}{\partial \dot{x}_k} = \lambda_k,$$

since the other coordinates, for $i \neq k$, are treated as constants. Therefore,

$$H = \sum_i \dot{x}_i \frac{\partial \mathcal{L}}{\partial \dot{x}_i} - \mathcal{L},$$
$$= \sum_i \dot{x}_i \lambda_i - \left[F + \sum_i \lambda_i (\dot{x}_i - f_i) \right],$$
$$= - \left[F - \sum_i \lambda_i f_i \right] = -H'.$$

In effect, we are choosing that control u that gives an extremal (maximum, minimum or point of inflection) of the time integral of the energy [69], namely

$$\int_{t=0}^{T} H' dt.$$

If this use of H' still looks rather like sleight-of-hand, there is an equivalent way of arriving at equation (31.4) that may be regarded as more consistent with the principles of the calculus of variations. This is to regard u as yet another coordinate, to be treated on the same footing as the x_i. So go back to the cost function J, and Lagrangian $\mathcal{L}$ from equation (31.1), and demand that

$$\frac{d}{dt}\left(\frac{\partial \mathcal{L}}{\partial \dot{u}}\right) = \frac{\partial \mathcal{L}}{\partial u}.$$

Since $\mathcal{L}$ is independent of $\dot{u}$, this reduces to

$$\frac{\partial \mathcal{L}}{\partial u} = 0.$$

But plugging in equation (31.1) and noting that the λ_i and $\dot{x}_i$ are independent of u,

$$0 = \frac{\partial \mathcal{L}}{\partial u} = \frac{\partial}{\partial u}\left[F(x_i, u, t) - \sum_i \lambda_i(t) f_i(x_j, u, t)\right],$$

$$= \frac{\partial H'}{\partial u}, \text{ from equation (31.3)},$$

and so equation (31.4) is retrieved.

31.2 Analysis of Moon Landing Problem

Returning to the moon-landing problem, we simplify the situation to one vertical dimension y, measured above the surface of the moon. Gravity is assumed constant, with acceleration g as before (but a different value to that on earth), so the dynamics of the landing craft are governed by the equation:

$$m(t)\frac{d^2 y}{dt^2} = -m(t)g + u(t), \tag{31.5}$$

where $u(t)$ is the rocket thrust and $m(t)$ is the mass of the entire landing craft, inclusive of fuel. Since the rockets are burning fuel, the mass m must vary with time, as indicated, and we assume that

$$\frac{dm}{dt} = -ku(t), \text{ for some constant } k > 0.$$

This just assumes that the rate of change of mass is proportional to the instantaneous rocket thrust. The thrust level is also obviously limited, so $0 \leq u(t) \leq U$ for some maximum value U.

Initially, at $t = t_0$, we assume that the landing craft is at some altitude y_0 above the surface and in free-fall downwards with a speed v_0 and mass m_0. The control u should be chosen so that the craft touches down at some later time T with zero speed.

The kinematics of the situation then reduce to the following set of first-order differential equations:

$$\frac{dy}{dt} = v,$$
$$\frac{dv}{dt} = -g + \frac{u}{m},$$
$$\frac{dm}{dt} = -ku.$$

The first two of these may be recombined to produce equation (31.5), so no magic has been involved in the translation. This scheme of adding an extra variable is a neat way of reducing a second-order differential equation to a pair of first-order ones. We now have three dynamic variables, y, v and m, plus the control u.

31.2. ANALYSIS OF MOON LANDING PROBLEM

The next requirement is to formulate the cost function — the basic quantity to be minimised. The fuel burned during the descent is given by

$$\int_{t_0}^{T} |\dot{m}|\, dt = k \int_{t_0}^{T} u(t)\, dt,$$

so the basic cost function will be:

$$J = \int_{t_0}^{T} u(t)\, dt,$$

since k is a constant and can be omitted.

Following the Pontryagin method from the preceding section, extend J to incorporate the above three kinematic equations:

$$J = \int_{t_0}^{T} \left[u(t) + \lambda_1(t)\left(\dot{y} - v\right) + \lambda_2(t)\left(\dot{v} + g - \frac{u(t)}{m(t)}\right) + \lambda_3(t)\left(\dot{m} + ku(t)\right) \right] dt,$$

from which

$$\mathcal{L} = u(t) + \lambda_1(t)\left(\dot{y} - v\right) + \lambda_2(t)\left(\dot{v} + g - \frac{u(t)}{m(t)}\right) + \lambda_3(t)\left(\dot{m} + ku(t)\right).$$

This is now in the form of equation (31.1), and we can make the following identifications:

$$F = u, \quad x_1 = y, \quad x_2 = v, \quad x_3 = m;$$
$$f_1 = v, \quad f_2 = -g + \frac{u}{m}, \quad f_3 = -ku.$$

Therefore, summarising the resulting 6 differential equations — 3 from the kinematic equations and 3 from equation (31.2) — we end up with:

$$\dot{y} = v,$$
$$\dot{v} = -g + \frac{u}{m},$$
$$\dot{m} = -ku,$$
$$\dot{\lambda}_1 = 0,$$
$$\dot{\lambda}_2 = -\lambda_1,$$
$$\dot{\lambda}_3 = \frac{\lambda_2 u}{m^2}.$$

The fourth and fifth of these may be solved immediately to give $\lambda_1 = a$ and $\lambda_2 = b - at$, where a and b are constants.

We also find that

$$H' = u - \lambda_1 v + \lambda_2\left(g - \frac{u}{m}\right) + \lambda_3 ku.$$

Rearranging this to separate out the control u then gives:

$$H' = \left(-\lambda_1 v + \lambda_2 g\right) + \left(1 - \frac{\lambda_2}{m} + \lambda_3 k\right) u.$$

Differentiating this with respect to u, namely seeking $\partial H / \partial u = 0$, gives no information about u itself. The situation can be clarified if H is written in the form $H = p + qu$, where the parameters p and q are independent of u. This is then just the equation for a straight line in u-space, for which there are no upper or lower limits — except those imposed by thruster limitations to $0 \leq u \leq U$. We conclude that in order to find the extremal of H, we must choose u to be one or other of its limiting values, namely $u = 0$ or $u = U$, or one after the other.

From the physics of the situation, then, choose $u = 0$ up to some time t_s, and thereafter use $u = U$ until touchdown. The alternative choice, with $u = U$ initially and then $u = 0$ after t_s, means that the craft will be accelerating downwards at impact, thus crashing into the surface.

This is as far as it is actually necessary to go from the optimal control point of view — the rest of the problem consists of solving the differential equations subject to the boundary conditions, and ensuring that the mass m exceeds some minimum value at touch-down[2]. In summary, then, we have the following pair of differential equations:

For $t < t_s$: $\dot{y} = v$, $\dot{v} = -g$, $\dot{m} = 0$,
and for $t \geq t_s$: $\dot{y} = v$, $\dot{v} = -g + U/m$, $\dot{m} = -kU$.

It can be assumed that k, g and U are known, while at the start of descent y, v and m are also provided. At touch-down, both y and v must be zero, with $m \geq \mu$ for some specified minimum mass μ. The 'switching time' t_s must be inferred, however, implying that iterative (numerical) solutions provide the best way forward.

As might have been inferred from the introductory words at the start of this chapter, exploration of optimal control theory received a significant spur during the 'space race' in the 1960s; the other contributory factor was the increasing availability of computer power. Rather more down to earth applications are not difficult to find, though; one example involves pushing a car into a garage, the car starting and ending at rest and the aim being to minimise the total time taken. It turns out that the control u in this case also needs to operate at its extremes — full force into the garage for a time and then full force in the opposite direction for the remainder of the interval [69].

---

[2] To ensure that enough fuel is available for the controlled descent and (if relevant) a return flight.

Chapter 32

Acknowledgements

THIS book would not have been written without encouragement from Jane and Iona, who have had to put up with my frequent absences in front of the computer, as well as occasional mental blankness due to equation overload.

It is doubtless unusual to acknowledge the contribution of politicians, but when I first went to university successive UK governments actually invested in education. Without a financial grant, my parents could not have afforded to send me to university, in which event it is unlikely that this book would have been written.

Looking back on it, the applied mathematics course that I took at the University of Wales was unusually wide in scope, covering several areas of mathematical physics in addition to the more 'traditional' material such as differential equations, complex analysis and so on. This breadth of background stood me in good stead throughout my working life.

I also gratefully acknowledge the advice and support of my advisor and professors at the Applied Mathematics Program, University of Washington, Seattle. Those several years of study in graduate school would not have been possible without funding from the US Air Force.

I have been reluctant to burden friends and relatives with proof-reading the text (not to mention the equations), and to the best of my knowledge there are no errors anywhere (I would say that, of course).

Bibliography

[1] *It All Adds Up — The Story of People and Mathematics*, M. Launay, William Collins, 2018.

[2] *The Road to Reality*, R. Penrose, Jonathan Cape, 2004.

[3] *Is God a Mathematician?*, M. Livio, Simon and Schuster, 2009.

[4] *Weapons of Math Destruction*, C. O'Neil, Penguin, 2016.

[5] *Calculating the Cosmos: How Mathematics Unveils the Universe*, I. Stewart, Profile Books 2017.

[6] *Advanced Engineering Mathematics*, E. Kreyszig, Sixth Edition, Wiley, 1988.

[7] *Mathematics from the Birth of Numbers*, J. Gullberg, W.W. Norton and Co., 1997.

[8] *Latitude: The Astonishing Adventure that Shaped the World*, N. Crane, Penguin Books, 2021.

[9] *Calculus*, T. Apostol, Volume 1, Second edition, Ginn-Blaisdell International Textbooks, 1967.

[10] *Extraterrestrial Cause for the Cretaceous-Tertiary Extinction: Experimental Results and Theoretical Interpretation*, L.W. Alvarez, W. Alvarez, F. Asaro and H.V. Michel, Science, Vol. 208, No. 4448, June 1980.

[11] *Science Data Book*, ed. R.M. Tennant, Oliver and Boyd, 1971.

[12] *When Life Nearly Died: The Greatest Mass Extinction of All Time*, M.J. Benton, Thames and Hudson, 2003.

[13] *Expected Number of Alignments in a Uniform Random Distribution*, P. Easthope, Studia z Automatyki i Informatyki, Vol. 43, 2018.

[14] *An Elementary Introduction to the Theory of Probability*, B.V. Gnedenko and A.Ya. Khinchin, Dover, 1962.

[15] *The Surprising Probability of Long Runs*, M.F. Schilling, Mathematics Magazine, Vol. 85, pages 141 to 149, 2012.

[16] *Humble Pi: A Comedy of Maths Errors*, M. Parker, Allen Lane, 2019.

[17] https://www.bmj.com/content/373/bmj.n1411/rr

[18] *Dictionary of Mathematics*, E.J. Borowski and J.M. Borwein, Second edition, Collins, 2002.

[19] https://www.lawteacher.net/free-law-essays/criminal-law/miscarriage-of-justice-case-sally-clark.php

[20] https://rss.org.uk/news-publication/news-publications/2022/section-group-reports/rss-publishes-report-on-dealing-with-uncertainty-i/

[21] https://www.cebm.ox.ac.uk/news/views/the-prosecutors-fallacy

[22] *One In Millions, Billions and Trillions: Lessons from People v. Collins (1968) for People v. Simpson (1995)*, J.J. Koehler, J. Legal Education, Vol. 47, No. 2, 1997.

[23] *Math On Trial: How Numbers Get Used and Abused in the Courtroom*, L. Schneps and C. Colmez, Basic Books, 2013.

[24] *DNA Evidence: Probability, Population Genetics and the Courts*, D.H. Kaye, Journal Articles, Volume 7, Penn State Law eLibrary, 1993.

[25] *The Next Nuclear Meltdown*, New York Times, 8 May 1985, A, page 26.

[26] *Statistical Review of Nuclear Power Accidents*, M. Hofert and M.V. Wüthrich, Asia Pacific Journal of Risk and Insurance, Vol. 7, Issue 1, 2013.

[27] https://world-nuclear.org/information-library/current-and-future-generation/nuclear-power-in-the-world-today.aspx

[28] *Weapon-Target Assignment Problem: Exact and Approximate Solution Algorithms*, A.C. Andersen, K. Pavlikov and T.A.M Toffolo, Annals of Operations Research, Vol. 312, No. 2, 2022.

[29] *Application of a Dynamic Programming Algorithm for Weapon Target Assignment*, L. Hammond, DST Group-TR-3221, Australian Dept. of Defence, 2016.

[30] *Weapon Target Assignment with Combinatorial Optimization Techniques*, A. Tokgöz and S. Bulkan, Int. J. Advanced Research in Artificial Intelligence, Vol. 2, No. 7, 2013.

[31] *Calculus Made Easy*, S. Thompson, Second Edition, MacMillan and Co., 1918.

[32] *Numerical Recipes in C. The Art of Scientific Computing*, W.H. Press, S.A. Teukolsky, W.T. Vetterling and B.P. Flannery, Second edition, Cambridge University Press, 1992.

BIBLIOGRAPHY

[33] *Mathematical Methods in Science and Engineering*, J. Heading, Edward Arnold (Publishers), 1963.

[34] *Applied Numerical Analysis*, C.F. Gerald, Addison-Wesley, 1978.

[35] *Perturbation Methods in Applied Mathematics*, J. Kevorkian and J.D. Cole, Springer-Verlag, 1981.

[36] *Table of Integrals, Series and Products*, I.S. Gradshteyn and I.M. Ryzhik, Academic Press, 2007.

[37] *Fluid Dynamics*, W.F. Hughes and J.A. Brighton, Schaum's Outline Series, McGraw-Hill, 1967.

[38] *Properties of Matter*, H.N.V. Temperley, University Tutorial Press, 1965.

[39] *The Role of Friction in the Static Equilibrium of a Fixed Ladder: Theoretical Analysis and Experimental Test*, M.P. Silverman, World Journal of Mechanics, Vol. 8, 2018. https://www.scirp.org/journal/wjm

[40] https://www.engineeringtoolbox.com/friction-coefficients-d_778.html

[41] *Statics of a Ladder Leaning Against a Rough Wall*, K. Mendelson, American Journal of Physics, Vol. 63, No. 2, 1995.

[42] https://ocw.tudelft.nl/wp-content/uploads/Bending-Deflection.pdf

[43] *The Hidden Reality: Parallel Universes and the Deep Laws of the Cosmos*, B. Greene, Penguin Books, 2012.

[44] *Vector Analysis*, M. Spiegel, Schaum's Outline Series, McGraw-Hill, 1959.

[45] *Detecting Periodic Patterns in Unevenly Spaced Gene Expression Time Series Using Lomb-Scargle Periodograms*, E.F. Glynn, J. Chen and A.R. Mushegian, Bioinformatics, Vol. 22, No. 3, 2006.

[46] *Understanding Quantum Mechanics*, G. Troup, Methuen, 1968.

[47] *Operational Resource Theory of Imaginarity*, K-D. Wu, T.V. Kondra, S. Rana, C.M. Scandolo, G-Y. Xiang, C-F. Li, G-C. Guo and A. Streltsov, Physical Review Letters, March 2021. DOI:10.1103/PhysRevLett.126.090401.

[48] *Complex Variables*, M. Spiegel, Schaum's Outline Series, McGraw-Hill, 1964.

[49] *Quaternions*, Y-B. Jia, Iowa State University, 2022
https://faculty.sites.iastate.edu/jia/files/inline-files/quaternion.pdf

[50] *Quaternions and Dynamics*, B. Graf, arXiv:0811.2889v1[math.DS], November 2008.

[51] *The Quaternion Group and Modern Physics*, P.R. Girard, European J. Physics, Vol. 5, No. 1, 1984.

[52] *Mathematical modelling of the spread of the coronavirus disease 2019 (COVID-19) taking into account the undetected infections*, B. Ivorra, M. Fernández, M. Vela-Pérez and A. Ramos, Commun. Nonlinear Sci. Numer. Simul., September 2020.

[53] *Reproduction Number (R) and Growth Rate (r) of the COVID-19 Epidemic in the UK*, Royal Society, August 2020 (`set-covid-19-R-estimates.pdf`)

[54] *Notes on R_0*, J.H. Jones, Dept. Anthropological Sciences, Stanford University, May 2007 (`Jones-on-R0.pdf`)

[55] *The Basic Reproduction Number (R_0) of Measles: a Systematic Review*, F.M. Guerra, et al, The Lancet Infectious Diseases, Vol. 17, Issue 12, December 2017

[56] https://www.vaccinestoday.eu/stories/what-is-r0/

[57] https://www.uspharmacist.com/article/measles-and-the-mmr-vaccine

[58] https://en.wikipedia.org/wiki/Herd_immunity

[59] *Operational Mathematics*, R.V. Churchill, McGraw-Hill Book Company, 1958.

[60] *The Radon Transform*, C. Høilund, Aalborg University, 2007. `misp.cs.cmu.edu/courses/fall2012/lectures/Carsten_Hoilund_Radon.pdf`

[61] *Inverting the Circular Radon Transform*, N.J. Redding and G.N. Newsam, Australian Defence Science and Technology Organisation, DSTO-RR-0211, 2001.

[62] https://thatsmaths.com/2013/03/07/ct-scans-and-the-radon-transform

[63] *Radon Spectrum and its Application for Small Moving Target Detection*, Y. Dong, Australian Defence Science and Technology Organisation, DSTO-TR-3103, 2015.

[64] *Applied Optimal Estimation*, A. Gelb (ed.), MIT Press, 1989.

[65] *A Tutorial on Particle Filtering and Smoothing: Fifteen Years Later*, A. Doucet and A.M. Johansen, 2012 https://www.stats.ox.ac.uk/ doucet/doucet_johansen_tutorialPF2011.pdf

[66] *Unscented Filtering and Nonlinear Estimation*, S.J. Julier and J.K. Uhlmann, Proceedings of the IEEE, Vol. 92, No. 3, March 2004.

[67] *The Fokker-Planck Equation: Methods of Solution and Applications*, H. Risken, Springer, 1996.

[68] https://sites.me.ucsb.edu/ moehlis/moehlis_papers/appendix.pdf

BIBLIOGRAPHY

[69] *Calculus of Variations*, J.W. Craggs, George Allen and Unwin, 1973.

[70] *Mechanics*, L.D. Landau and E.M. Lifshitz, Volume 1 in Course of Theoretical Physics, Pergamon Press, 1976.

[71] *On the Zero-Energy Universe*, M. Berman, `arXiv:gr-qc/0605063v3`, August 2009.

[72] *The Transactional Interpretation of Quantum Mechanics and Quantum Nonlocality*, J. Cramer, arXiv:1503.00039v1 [quant-ph], February 2015.

[73] *Chaos*, J. Gleick, Sphere Books, 1987.

[74] *Deterministic Nonperiodic Flow*, E.N. Lorenz, J. Atmospheric Sciences, Vol. 20, No. 2, 1963.

[75] *A Chaotic Attractor from Chua's Circuit*, T. Matsumoto, IEEE Trans. on Circuits and Systems, Vol. CAS-31, No. 12, 1984.

[76] *A Numerical Experiment on the Chaotic Behaviour of the Solar System*, J. Laskar, Nature, Vol. 338, 1989.

[77] *The Origin of Chaos in the Solar System through Computer Algebra*, F. Mogavero and J. Laskar, Astronomy and Astrophysics, Vol. 662, L3, 2022.

[78] *Chaotic Water Drop Experiment*, S. Errede, Physics 403 Lab., Physics Department, University of Illinois, 2004.

[79] *Chaotic Behaviour of the Earth System in the Anthropocene*, A.E. Bernadini, O. Bertolami and F. Francisco, arXiv:2204.08955v2 [astro-ph.EP] 21 Apr 2022.

[80] *Chaos and Weather Prediction*, R. Buizza, Meteorological Training Course Lecture Series, European Centre for Medium-Range Weather, 2002.

[81] *The Chaotic Behaviour of the Spread of Infection During the COVID-19 Pandemic in the United States and Globally*, N. Sapkota, W. Karwowski, M.R. Davahli, A. Al-Juaid, R. Taiar, A. Murata, G. Wróbel and T. Marek, IEEE Access, DOI 10.1109/ACCESS.2021.3085240.

[82] *Chaos in Electronic Circuits*, T. Matsumoto, Proc. IEEE, Vol. 75, No. 8, 1987.

[83] `https://en.wikipedia.org/wiki/List_of_tsunamis`

[84] *Lecture 8: The Shallow-Water Equations*, H. Segur, 2009.
`https://gfd.whoi.edu/wp-content/uploads/sites/18/2018/03/lecture8-harvey_136564.pdf`

[85] *On The Electrodynamics of Moving Bodies*, A. Einstein, Annalen der Physik, 17, 1905.

[86] *An Introduction to Tensor Calculus and Relativity*, D.F. Lawden, Chapman and Hall, 1975.

[87] *Observer with a Constant Proper Acceleration*, C. Semay, arXiv:physics/0601179v1 [physics.ed-ph], 2006.

[88] *Lectures on Dynamics and Relativity*, D. Tong, University of Cambridge, 2013. https://www.damtp.cam.ac.uk/user/tong/relativity/seven.pdf

[89] *The Meaning Of Relativity*, A. Einstein, Chapman and Hall, 1973.

[90] *Introduction to General Relativity*, R.J. Adler, M.J. Bazin and M. Schiffer, McGraw-Hill, 1975.

[91] *Schrödinger's Kittens and the Search for Reality*, J. Gribbin, Phoenix, 1997.

[92] *Relative-Distance Machian Theories*, J.B. Barbour, Nature, Vol. 249, No. 5455, May 1974.

[93] *The Equivalence Principle, Uniformly Accelerated Reference Frames, and the Uniform Gravitational Field*, G. Muñoz and P. Jones, American J. Physics, Vol. 78, Section 4, 2010.

[94] *Constant Acceleration and the Equivalence Principle.* K. McDonald, Joseph Henry Labs, Princeton University, May 2022.

[95] *Black Holes: The Key to Understanding the Universe*, B. Cox and J. Forshaw, William Collins, 2022.

[96] *Relativistic Effects on Clocks Aboard GPS Satellites*, C.E. Mungan, The Physics Teacher, Vol. 44, October 2006.

[97] www.astronomy.ohio-state.edu/~pogge/Ast162/Unit5/gps.html

[98] *The World According to Physics*, J. Al-Khalili, Princeton University Press, 2020.

[99] https://mason.gmu.edu/~treid5/Math447/GPSEquations

[100] *Relativity in the Global Positioning System*, N. Ashby, Living Reviews in Relativity, Max Planck Institute for Gravitational Physics, January 2003. https://link.springer.com/article/10.12942/lrr-2003-1

[101] *How to Derive the Schrödinger Equation*, D.W. Ward and S. Volkmer, American J. Physics, October 2006 arXiv:physics/0610121v1

[102] *Basic Quantum Mechanics*, K. Ziock, Wiley, 1969.

[103] *Quantum Mechanics*, L.D. Landau and E.M. Lifshitz, Volume 2 in Course of Theoretical Physics, Pergamon Press, 1974.

[104] *Reality Is Not What It Seems: The Journey to Quantum Gravity*, C. Rovelli, Penguin Books, 2016.

[105] *The Elegant Universe*, B. Greene, Vintage, 2000.

[106] *The Spell of Mathematics*, W.J. Reichmann, Pelican Books, 1972.

[107] *Mathematics for the Million*, L. Hogben, George Allen and Unwin, 1942.

[108] *Astronomy and Mathematics in Ancient China: the Zhou bi suan jing*, C. Cullen, Cambridge University Press, 1996.

[109] *Vector: A Surprising Story of Space, Time, and Mathematical Transformation*, R. Arianrhod, University of Chicago Press, 2024.

[110] *Introduction to Mathematical Statistics*, R.V. Hogg, J.W. McKean and A.T. Craig, Pearson, 2013.

[111] *Probability of Track Impact in Defended Area: Use of Green's Theorem in the Plane*, P. Easthope, Studia z Automatyki i Informatyki, Vol. 43, 2018.

[112] *Introduction to Probability Models*, S.M. Ross, Academic Press, 2007.

Appendix A

Pythagoras' Theorem

THE theorem ascribed to him was known to the Babylonians around 2000 BC. Numerous proofs exist [7], [2], [106], but one of the simplest to comprehend comes from an ancient Chinese mathematical text, the Book of Chou Pei Suan Ching 周髀算經[1] [107], [108]; see also [109]. The logic is illustrated in Figure A.1, and consists of two nested squares.

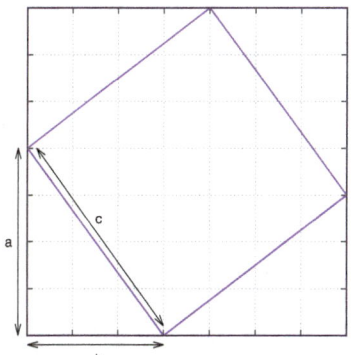

Figure A.1: Proof of Pythagoras' theorem

The right-angled triangle of interest is labelled with sides a, b and c. We work on the basis of areas, with the understanding that the area A of the outer square must be equal to that of the inner square, B, plus the areas of the four triangles, each with area Δ. That is,

$$A = B + 4\Delta.$$

---

[1] Zhou Bi Suan Jing, in Pinyin. The original text dates back to the Zhou dynasty, 1046 to 256 BC, and was augmented in the Han dynasty, 202 BC to 220 AD. It is not entirely clear, though, that the original Chinese text was written so as to constitute a proof of the Pythagorean theorem in the current sense of the word.

The three areas are given by:

$A = (a+b)^2$; from symmetry, the length of each side must be $a+b$,

$B = c^2$,

$\Delta = \dfrac{1}{2}ab$. This is half of the area of the rectangle with sides a and b.

Putting the bits together, we get:

$$(a+b)^2 = c^2 + 2ab,$$

which simplifies to

$$a^2 + b^2 = c^2.$$

The symmetry in the situation is highlighted by the use of smaller grid squares, as shown in Figure A.1. These are reminiscent of square tiles filling the outer area, which may be how the theorem was uncovered in the first place. The whole diagram also has rotational symmetry about its centre point, in integer multiples of $90°$ (or $\pi/2$).

Appendix B

Medical Test Probabilities

SECTION 6.2 presented an example set of probabilities associated with a medical test for disease X (entirely hypothetical). The background rate of this disease in the general population is defined as $P(X) = 0.01$. The medical test in question has a missed detection rate of $P(-|X) = 0.05$ and a false alarm rate of $P(+|H) = 0.03$; (these are the two failure rates for the test). See Figure 6.2 for the sample numbers associated with these probabilities and Section 6.2 for an explanation of the notation used.

Three intermediate derived numbers are required:

- $P(H) = 1 - P(X) = 0.99$ (which just means that 99% of people are free of disease X).

- $P(+|X) = 1 - P(-|X) = 0.95$, the probability that a person with disease X will correctly end up with a positive test.

- $P(-|H) = 1 - P(+|H) = 0.97$, the probability that a healthy person will, as expected, test negative.

The four derived probabilities and their values are then as follows:

$$P(X|+) = \frac{P(+|X)P(X)}{P(+|X)P(X) + P(+|H)P(H)} = 0.242347,$$

$$P(X|-) = \frac{P(-|X)P(X)}{P(-|X)P(X) + P(-|H)P(H)} = 0.00052,$$

$$P(H|+) = \frac{P(+|H)P(H)}{P(+|H)P(H) + P(+|X)P(X)} = 1 - P(X|+) = 0.757653,$$

$$P(H|-) = \frac{P(-|H)P(H)}{P(-|H)P(H) + P(-|X)P(X)} = 1 - P(X|-) = 0.99948.$$

We also have $P(+) = P(+|X)P(X) + P(+|H)P(H) = 0.0392$, and $P(-) = P(-|X)P(X) + P(-|H)P(H) = 1 - P(+) = 0.9608$; these are the denominators in the above equations.

Appendix C

Expression for Pendulum Period

COMMENCE from equation (11.12),

$$\frac{d^2\theta}{dt^2} = -\frac{g}{l}\sin\theta.$$

Define $u = d\theta/dt$, so

$$\frac{du}{dt} = -\frac{g}{l}\sin\theta.$$

Write

$$\frac{du}{dt} = \frac{du}{d\theta}\frac{d\theta}{dt},$$
$$= u\frac{du}{d\theta}.$$

(This is a useful technique for solving second-order differential equations, especially nonlinear ones.)

Therefore,

$$u\frac{du}{d\theta} = -\frac{g}{l}\sin\theta,$$

thereby bypassing the need to worry about the time coordinate, at least for the time being. This is readily integrated:

$$\frac{1}{2}u^2 + C = \frac{g}{l}\cos\theta,$$

including the usual constant of integration.

Suppose now that the pendulum commences, at $t = 0$, with $\theta = -\theta_0$ and $d\theta/dt = u = 0$. This gives

$$C = \frac{g}{l}\cos\theta_0,$$

so

$$\frac{1}{2}u^2 = \frac{g}{l}\{\cos\theta - \cos\theta_0\}.$$

Reverting to $d\theta/dt$,

$$\frac{d\theta}{dt} = \sqrt{\frac{2g}{l}}\sqrt{\cos\theta - \cos\theta_0},$$

in which the positive root is selected, leading to

$$t\sqrt{\frac{2g}{l}} = \int_{-\theta_0} \frac{d\theta}{\sqrt{\cos\theta - \cos\theta_0}}.$$

Let the period be T, as before. Then *half* of one period will be given by the motion between $\theta = -\theta_0$ and $\theta = \theta_0$:

$$\frac{T}{2}\sqrt{\frac{2g}{l}} = \int_{-\theta_0}^{\theta_0} \frac{d\theta}{\sqrt{\cos\theta - \cos\theta_0}}. \tag{C.1}$$

This *can* be integrated numerically as it stands, albeit with some care needed at the two end points to avoid infinities occurring when $\cos\theta = \cos\theta_0$.

A better approach, though, can be obtained by changing the integration variable. First note that the integrand is symmetric either side of $\theta = 0$, so that:

$$\frac{T}{2}\sqrt{\frac{2g}{l}} = 2\int_{0}^{\theta_0} \frac{d\theta}{\sqrt{\cos\theta - \cos\theta_0}}.$$

Then define

$$s = \tan(\theta/2) \text{ with } s_0 = \tan(\theta_0/2),$$

which allows us to use the half-angle formula[1],

$$\cos\theta = \frac{1-s^2}{1+s^2}. \tag{C.2}$$

As a brief diversion, showing that this equation holds is quite straightforward; insert the definition of s into equation (C.2) to get:

$$\cos\theta = \frac{1 - \tan^2(\theta/2)}{1 + \tan^2(\theta/2)},$$

$$= \frac{1 - \frac{\sin^2(\theta/2)}{\cos^2(\theta/2)}}{1 + \frac{\sin^2(\theta/2)}{\cos^2(\theta/2)}}, \text{ from the definition of } \tan x = \sin x/\cos x,$$

$$= \frac{\cos^2(\theta/2) - \sin^2(\theta/2)}{\cos^2(\theta/2) + \sin^2(\theta/2)},$$

$$= \cos^2(\theta/2) - \sin^2(\theta/2), \text{ since } \cos^2 x + \sin^2 x = 1.$$

---

[1] There are equivalents for $\sin\theta$ and $\tan\theta$.

But, from equation (3.6),
$$\cos(\theta/2 + \theta/2) = \cos^2(\theta/2) - \sin^2(\theta/2),$$
and so the identity is proved.

Now differentiate $s = \tan(\theta/2)$ to get[2]
$$ds = \frac{d}{d\theta}\tan(\theta/2)\, d\theta = \frac{1}{2}\sec^2(\theta/2)\, d\theta = \frac{1}{2}(1+s^2)\, d\theta,$$
so that
$$d\theta = \frac{2\, ds}{1+s^2}.$$

Putting this lot together,
$$\frac{T}{2}\sqrt{\frac{2g}{l}} = 2\int_{s=0}^{s_0}\left[\left(\frac{1-s^2}{1+s^2}\right) - \left(\frac{1-s_0^2}{1+s_0^2}\right)\right]^{-1/2}\frac{2\, ds}{1+s^2},$$
$$= 4\int_{s=0}^{s_0}\left[\frac{(1-s^2)(1+s_0^2) - (1+s^2)(1-s_0^2)}{(1+s^2)(1+s_0^2)}\right]^{-1/2}\frac{ds}{1+s^2},$$
$$= 4\int_{s=0}^{s_0}\left[\frac{(1+s^2)(1+s_0^2)}{2(s_0^2-s^2)}\right]^{1/2}\frac{ds}{1+s^2},$$
$$= \frac{4\sqrt{1+s_0^2}}{\sqrt{2}}\int_{s=0}^{s_0}\frac{ds}{\sqrt{1+s^2}\sqrt{s_0^2-s^2}}.$$

This can be further simplified by writing $s = s_0 \sin\psi$, which results in
$$\frac{T}{2}\sqrt{\frac{2g}{l}} = \frac{4\sqrt{1+s_0^2}}{\sqrt{2}}\int_{\psi=0}^{\pi/2}\frac{d\psi}{\sqrt{1+s_0^2\sin^2\psi}},$$
and so
$$T = 4\sqrt{\frac{l}{g}}\sqrt{1+s_0^2}\int_{\psi=0}^{\pi/2}\frac{d\psi}{\sqrt{1+s_0^2\sin^2\psi}}, \tag{C.3}$$

recalling that $s_0 = \tan(\theta_0/2)$. If θ_0 is small, so too will be s_0, and the familiar approximate equation $T \approx 2\pi\sqrt{l/g}$ is retrieved.

In contrast to the integral in equation (C.1), the integrand here is well-behaved at both $\psi = 0$ and $\pi/2$, and numerical integration presents no problems.

An approximation to $O(s_0^4)$ can be obtained from equation (C.3) by expanding both the integrand and $\sqrt{1+s_0^2}$ in powers of s_0^2, integrating term by term and dropping terms of order s_0^6 and higher, resulting in the following:
$$T \approx 2\pi\sqrt{\frac{l}{g}}\left(1 + \frac{1}{4}s_0^2 - \frac{7}{64}s_0^4 + O(s_0^6)\right).$$

---

[2] To differentiate $\tan x$ with respect to x, express $\tan x = \sin x/\cos x$ and then use equation (7.9).

Appendix D

Solution for Heavy Chain

REFERRING back to Section 23, the shape adopted by a heavy uniform chain will be given by the equation:

$$\frac{d}{dx}\left(\frac{yy'}{\sqrt{1+y'^2}}\right) = \sqrt{1+y'^2}.$$

To aid in solving this, it can be seen that x is not explicitly present on either side of the equation, so replace the derivative with respect to x with a derivative with respect to y:

$$\frac{d}{dx} = \frac{dy}{dx}\frac{d}{dy} = y'\frac{d}{dy}.$$

Thus,

$$y'\frac{d}{dy}\left(\frac{yy'}{\sqrt{1+y'^2}}\right) = \sqrt{1+y'^2} \;\Rightarrow\; \frac{d}{dy}\left(\frac{yy'}{\sqrt{1+y'^2}}\right) = \frac{\sqrt{1+y'^2}}{y'},$$

after a bit of rearrangement. There is a pattern visible here, in that the quantity on the right-hand-side appears in inverse form under the derivative on the left. So define a new intermediate variable z, with

$$z = \frac{y'}{\sqrt{1+y'^2}},$$

which gives

$$\frac{d}{dy}(yz) = \frac{1}{z}, \;\text{ which implies that }\; y\frac{dz}{dy} = \frac{1-z^2}{z}.$$

This separates out into the integral form:

$$\int \frac{dy}{y} = \int \frac{z\,dz}{1-z^2}.$$

APPENDIX D. SOLUTION FOR HEAVY CHAIN

Carrying out the integration:

$$\log_e y + \log_e C = -\frac{1}{2}\log_e(1 - z^2), \quad \Rightarrow \quad Cy = \frac{1}{\sqrt{1-z^2}},$$

where C is a constant. Square both sides and invert:

$$\frac{1}{C^2 y^2} = 1 - z^2 = \frac{1}{1 + y'^2}, \quad \text{after substituting the definition of } z.$$

After a bit more shuffling of terms,

$$\frac{dy}{dx} = \pm\sqrt{C^2 y^2 - 1}. \tag{D.1}$$

Now bring in the definitions of the hyperbolic sine and cosine functions:

$$\sinh x = \frac{1}{2}\left(e^x - e^{-x}\right),$$

$$\cosh x = \frac{1}{2}\left(e^x + e^{-x}\right),$$

the h in the name standing for 'hyperbolic'. By inspection, it can be seen that these are derivatives of one another, so if

$$y = \frac{\pm 1}{C}\cosh\left(Cx - D\right),$$

equation (D.1) is satisfied. Here, D is another constant.

Appendix E

Solution of Quadratic Equations

UPPOSE we have an algebraic equation of the form:
$$ax^2 + bx + c = 0,$$
to be solved for x.

Assume that a is not zero (if it is, then the equation simplifies to $bx + c = 0$, which is easy to solve). So divide throughout by a:
$$x^2 + \frac{b}{a}x + \frac{c}{a} = 0. \tag{E.1}$$

Now complete the square:
$$\left(x + \frac{b}{2a}\right)^2 - \frac{b^2}{4a^2} + \frac{c}{a} = 0.$$

To check, if the bracketed term is expanded out, the first two terms in equation (E.1) are retrieved, while the $b^2/4a^2$ terms cancel. So this equation is actually the same as equation (E.1).

Put the bracketed bit on its own and shift the other two bits over on the right:
$$\left(x + \frac{b}{2a}\right)^2 = \frac{b^2}{4a^2} - \frac{c}{a},$$
$$= \frac{1}{4a^2}\left(b^2 - 4ac\right), \text{ after a bit of rearrangement.}$$

Take the square root of both sides:
$$x + \frac{b}{2a} = \frac{\pm 1}{2a}\sqrt{b^2 - 4ac}, \text{ since } (\pm 1)^2 = 1.$$

Rearrange this to get the general solution:
$$x = \frac{1}{2a}\left(-b \pm \sqrt{b^2 - 4ac}\right).$$

Appendix F

Probability Density Mappings

THE majority of 'canned' random number generators, such as the base set provided in the C computer language, just give a uniformly-distributed integer somewhere between zero and a large value. 'Uniformly-distributed' means equi-probable anywhere in the range, whereas there is often a need to generate random numbers that are more concentrated in one region or another. So it is useful to have some means by which uniform random numbers may be converted into samples from other distributions, and this section gives a brief overview as to how this can be done. The exponential distribution is used as an example, which has probability density defined by:

$$p(y) = \frac{1}{\mu} e^{-y/\mu}, \text{ for } y \geq 0. \tag{F.1}$$

This density means that smaller values of y are more probable, with larger values being progressively rarer. It is a continuous distribution, so that the probability of finding a value between y and $y + dy$ is given by $p(y)dy$. The density has the following properties:

$$\int_{y=0}^{\infty} p(y)\, dy = 1 \text{ (normalisation)},$$

$$\int_{y=0}^{\infty} y\, p(y)\, dy = \mu, \text{ the mean value of } y.$$

The mapping of probability distributions is most systematically carried out by means of the *cumulative distribution*, defined as:

$$P(y \leq Y) = \int_{y=0}^{Y} p(y)\, dy.$$

This gives the probability of finding a value of y in the range 0 (or, more generally, from whatever the lower limit is) up to Y. Suppose now that we can find some equation linking a new variable x to y, so that

$$P(y \leq Y) = \int_{y=0}^{Y} p(y)\, dy \quad \text{(by definition)},$$

$$= \int_{x=x_0}^{X} \left[p(y)\frac{dy}{dx} \right] dx, \quad \text{for some as-yet unknown limits on } x,$$

$$= \int_{x=x_0}^{X} q(x)\, dx, \quad \text{defining } q(x) = p(y)\frac{dy}{dx},$$

$$= P(x \leq X), \quad \text{also by definition.}$$

If, then, $q(x)$ is a known probability density, it is possible to work out how x and y should be related, given $p(y)$ (see also [32] and [110]). In the present case, $q(x)$ represents a uniform distribution, so assume that this is defined from zero to unity, giving $q(x) = 1$. Therefore, since $p(y)$ is given by equation (F.1), we must have:

$$\frac{1}{\mu} e^{-y/\mu} \frac{dy}{dx} = 1.$$

Integrating this,

$$x - C = \int \frac{1}{\mu} e^{-y/\mu}\, dy = -e^{-y/\mu},$$

for some constant C. Rearranging,

$$y = -\mu \log_e (C - x).$$

We can then put $C = 1$, and since $1 - x$ gives the same random set as x by itself, we may as well use

$$y = -\mu \log_e x,$$

which is the required mapping (at least provided $x \neq 0$, which needs to be guarded against in software).

Appendix G

Green's Theorem in the Plane

THE general form of Green's theorem in the plane is as follows [6]:

$$\iint_R \left(\frac{\partial F_2}{\partial x} - \frac{\partial F_1}{\partial y} \right) dx\,dy = \oint_C (F_1\,dx + F_2\,dy),$$

where R denotes a closed bounded region in the x-y plane, having boundary C. The above equation relates a double integral over an area to a single integral over its contour boundary, the latter being carried out (by convention) in an anti-clockwise direction. Functions $F_1(x,y)$ and $F_2(x,y)$ are required to be continuous and to have continuous first derivatives $\partial F_1/\partial y$ and $\partial F_2/\partial x$ everywhere in R.

So how do we get the area A in equation (13.1) from Green's theorem? The area itself is, in double integral form, just

$$A = \iint_R dx\,dy.$$

Then work in two steps: first set $F_1 = 0$ and $F_2 = x$; then set $F_1 = -y$ and $F_2 = 0$. Combining the two outputs gives equation (13.1).

Besides enabling the calculation of complicated areas, as in Section 13.1, the theorem also has application in determining the probability of intersection of a Gaussian distribution with some closed planar domain of interest [111].

Appendix H

Inverse Laplace Transform

I have no memory of ever seeing the derivation of the inverse Laplace Transform when I was a student, although one or other lecture must have gone over it. It's not as complicated as I had anticipated when planning this section, as you will see, although I have only provided an outline derivation here.

A specific forward Laplace Transform operation will be needed during the analysis, that of the simple exponential $e^{\mu t}$, so that

$$\int_{t=0}^{\infty} e^{\mu t} e^{-st} dt = \int_{t=0}^{\infty} e^{-(s-\mu)t} dt = \left[\frac{e^{-(s-\mu)t}}{-(s-\mu)} \right]_{t=0}^{\infty} = \frac{1}{s-\mu}, \quad (H.1)$$

here assuming that $s > \mu$ in order that the exponential should go to zero as $t \to \infty$. Keep this equation in mind for later use.

Now go back to Cauchy's Integral Formula, equation (17.1), in which both s and z are to be regarded as complex variables:

$$f(s) = \frac{1}{2\pi i} \oint_C \frac{f(z) dz}{z - s},$$

the closed contour C being such as to enclose the point s.

With a suitable choice of C and some constraints on the form of $f(z)$ to ensure convergence of the integral, which requires that (at least)

$$f(z) \sim \frac{1}{|z|^k} \text{ for } k > 0 \text{ as } |z| \to \infty \text{ [59]},$$

it is possible to replace the closed contour C with a single vertical path from $\gamma - i\beta$ to $\gamma + i\beta$ in complex space. Therefore, under these conditions,

$$f(s) = \lim_{\beta \to \infty} \frac{-1}{2\pi i} \int_{\gamma-i\beta}^{\gamma+i\beta} \frac{f(z) dz}{z - s} = \lim_{\beta \to \infty} \frac{1}{2\pi i} \int_{\gamma-i\beta}^{\gamma+i\beta} \frac{f(z) dz}{s - z},$$

in which the real quantity γ is fixed[1]. It may help to clarify the next step by

---
[1] See [59] for the specific contour used here.

writing this slightly differently as:

$$f(s) = \lim_{\beta \to \infty} \frac{1}{2\pi i} \int_{\gamma-i\beta}^{\gamma+i\beta} f(z) \frac{1}{s-z} dz, \qquad (\text{H.2})$$

which just separates out the $1/(s-z)$ factor.

Now borrow equation (H.1) and replace μ with z, so that

$$\frac{1}{s-z} = \int_{t=0}^{\infty} e^{zt} e^{-st} dt.$$

Substitute the right-hand-side of this into equation (H.2) to replace the $1/(s-z)$ factor, giving:

$$f(s) = \lim_{\beta \to \infty} \frac{1}{2\pi i} \int_{z=\gamma-i\beta}^{\gamma+i\beta} f(z) \int_{t=0}^{\infty} e^{zt} e^{-st} dt\, dz,$$

(adding z to the lower limit in the outer integral to make it quite clear which limits apply to which integral).

Reverse the order of the integrations and add a pair of brackets to highlight the chunk of interest:

$$f(s) = \int_{t=0}^{\infty} e^{-st} \left[\lim_{\beta \to \infty} \frac{1}{2\pi i} \int_{z=\gamma-i\beta}^{\gamma+i\beta} e^{zt} f(z)\, dz \right] dt.$$

What this says, using words, is that the Laplace Transform of the square-bracketed quantity (which is a function of t only) results in $f(s)$. Therefore, the inverse transformation of $f(s)$ must equal to the square-bracketed quantity.

In symbols, if we use $\mathcal{L}$ to denote the forward Laplace Transform and $\mathcal{L}^{-1}$ to stand for the inverse operation, then

$$\mathcal{L}^{-1}(f(s)) = \lim_{\beta \to \infty} \frac{1}{2\pi i} \int_{z=\gamma-i\beta}^{\gamma+i\beta} e^{tz} f(z)\, dz,$$

which needs to be evaluated in the complex z-space ([48]).

Appendix I

Projection Slice Theorem

THIS appendix provides an outline derivation of the inverse Radon Transform, namely equation (20.6) in Section 20.2. To reiterate, we have the forward Radon Transform as

$$R(\rho, \theta) = \int_s f(\rho \cos\theta - s \sin\theta, \rho \sin\theta + s \cos\theta) \, ds, \tag{I.1}$$

integrating over the path length s along one X-ray beam.

We also need definitions for the unbounded forward and reverse Fourier Transforms, which are as follows [59]:

$$F(\mu) = \int_{x=-\infty}^{\infty} f(x) \, e^{i\mu x} \, dx, \text{ the transform into frequency } \mu \text{ space,}$$

$$f(x) = \frac{1}{2\pi} \int_{\mu=-\infty}^{\infty} F(\mu) \, e^{-i\mu x} \, d\mu; \text{ the inverse transformation.}$$

These follow as a generalisation into the complex domain from the sine and cosine transforms mentioned in Section 16. Care is needed to ensure that the integrals exist, which in turn imposes conditions on the possible forms of $f(x)$ and $F(\nu)$ (but we don't worry about that here).

Next define $\mu = 2\pi\nu$, to be consistent with the notation in Section 20.2. This results in:

$$F(\nu) = \int_{x=-\infty}^{\infty} f(x) \, e^{2\pi i \nu x} \, dx, \text{ with } f(x) = \int_{\nu=-\infty}^{\infty} F(\nu) \, e^{-2\pi i \nu x} \, d\nu.$$

To proceed, first take the forward one-dimensional Fourier Transform of $R(\rho, \theta)$

with respect to ρ:

$$F_1(\nu, \theta) = \int_{\rho=-\infty}^{\infty} R(\rho, \theta) e^{2\pi i \rho \nu} d\rho, \qquad (I.2)$$

$$= \int_\rho \int_s f(\rho \cos \theta - s \sin \theta, \rho \sin \theta + s \cos \theta) e^{2\pi i \rho \nu} ds\, d\rho,$$

after substituting equation (I.1) for $R(\rho, \theta)$. The subscript $_1$ in F_1 is just used to highlight that the first integral here is a one-dimensional transform; the integral limits are left implicit.

This is now a double integral in ρ-s space, with θ as a parameter. Shift back into the original Cartesian x-y coordinates using

$$x = \rho \cos \theta - s \sin \theta, \quad y = \rho \sin \theta + s \cos \theta,$$

which corresponds to a rotation of the coordinate system by an angle θ, so we can write:

$$F_1(\nu, \theta) = \int_x \int_y f(x, y) e^{2\pi i \nu (x \cos \theta + y \sin \theta)} dx\, dy. \qquad (I.3)$$

(See equation (20.4) in Section 20.2 to obtain ρ in terms of x and y).

Hold equation (I.3) in mind for the time being.

Now look at the two-dimensional Fourier Transform of $f(x, y)$ into p-q frequency space, defining

$$F_2(p, q) = \int_x \int_y f(x, y) e^{2\pi i p x} e^{2\pi i q y} dx\, dy,$$

$$= \int_x \int_y f(x, y) e^{2\pi i (px + qy)} dx\, dy, \qquad (I.4)$$

here combining the two exponential powers into a single expression. This looks suspiciously like equation (I.3) and the similarity can be made exact if we identify p with $\nu \cos \theta$ and q with $\nu \sin \theta$, so that

$$F_2(\nu \cos \theta, \nu \sin \theta) = \int_x \int_y f(x, y) e^{2\pi i \nu (x \cos \theta + y \sin \theta)} dx\, dy.$$

What the above lot of operations comes to is that it is possible to identify the one-dimensional Fourier Transform of $R(\rho, \theta)$ in equation (I.2) with the two-dimensional Fourier transform of $f(x, y)$. That is,

$$F_2(\nu \cos \theta, \nu \sin \theta) = F_1(\nu, \theta),$$

which is referred to as the *Projection Slice Theorem*. This, in turn, gives us a way of deriving $f(x, y)$ when we are given $R(\rho, \theta)$. To see this, invert equation (I.4), sticking with the p and q coordinates for a moment:

$$f(x, y) = \int_p \int_q F_2(p, q) e^{-2\pi i (px + qy)} dp\, dq.$$

In this double integral, change from Cartesian p-q coordinates into polar ν-θ coordinates using $p = \nu \cos\theta$, $q = \nu \sin\theta$, so that

$$f(x,y) = \int_\nu \int_\theta |\nu| F_2(\nu \cos\theta, \nu \sin\theta) e^{-2\pi i \nu (x \cos\theta + y \sin\theta)} d\theta\, d\nu.$$

The appearance of the $|\nu|$ factor is a consequence of the change from Cartesian to polar coordinates in the area integral; I don't intend to go any further into the derivation of determinants of Jacobian matrices, *etc*, but the details of the mapping can be found in [6].

Now reverse the order of the integrals over ν and θ, add back the appropriate integration limits and substitute equation (I.2) in place of $F_2(\nu \cos\theta, \nu \sin\theta)$, resulting in:

$$f(x,y) = \int_{\theta=0}^{\pi} \int_{\nu=-\infty}^{\infty} |\nu| \left\{ \int_{\rho=-\infty}^{\infty} R(\rho, \theta)\, e^{2\pi i \rho \nu}\, d\rho \right\} e^{-2\pi i \nu (x \cos\theta + y \sin\theta)}\, d\nu\, d\theta,$$

which is the filtered back-projection of the Radon Transform, equation (20.6) in Section 20.2.

I admit that the above derivation is a bit convoluted and it took me a while to figure it out (even with the assistance of some biscuits). But it's worth it.

Appendix J

Random Walk Probabilities

IN Section 22, we investigated the expected behaviour of a mosquito undergoing a random walk, in which the magnitude and direction of every step is random and independent of every other step. As a consequence of the analysis, it is expected that the mosquito will remain within the vicinity of its starting point, but that the uncertainty (standard deviation) s_n associated with its location will grow according to $s_n = \sigma\sqrt{n}$, where $\sigma = 0.6$ m is the mean step length and n is the number of steps taken.

It is useful to be able to derive the probability of finding the mosquito within a circle of radius R around its starting point (the origin in this case). To obtain this probability, we need the individual probability distributions associated with x_n and y_n; from the foregoing results, these are expected to be zero-mean and similarly distributed, so call the individual distributions $p(x)$ and $p(y)$ with $p(.)$ being the same function for both.

How to determine $p(.)$? This is thankfully simpler than might be apparent, since we can make use of the *central limit theorem* [112], which states that a sum of identically distributed random variables tends, in the limit as $n \to \infty$, to the normal distribution.

So we expect that the zero-mean distribution will be approximated by the following equation:

$$p(x) = \frac{1}{s_n\sqrt{2\pi}} e^{-x^2/2s_n^2}, \text{ with } s_n \text{ as above.} \tag{J.1}$$

And we can assume that $p(y)$ will take the same form.

The assumption that the distribution of x_n will be close to the normal distribution can be tested using the 5000-mozzie simulation data in Section 22. Take the set of 5000 values comprising the x_{600} samples (the final location for each simulated mosquito) and divide the minimum and maximum numeric span of these values into 40, say, discrete bins[1]. Then assign each x_{600} value into the appropriate bin and count the number in each bin; this is essentially creating a histogram and results in Figure J.1.

---

[1] Choosing the appropriate number of bins requires a compromise between smoothness and resolution. Trial and error is usually required.

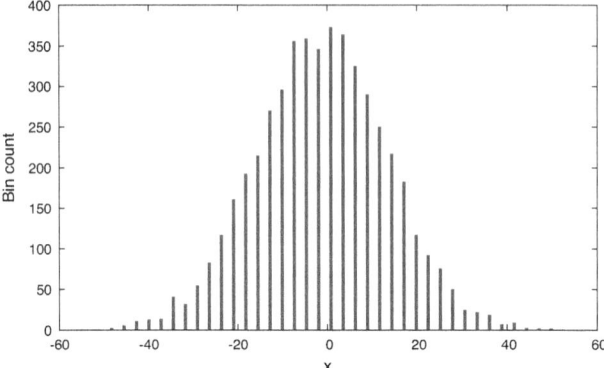

Figure J.1: Bin counts for x at $n = 600$

This picture at least *looks* Gaussian in form, being concentrated near the origin and falling off quasi-exponentially, but is the histogram data consistent with the expected normal distribution? To check out this question, the corresponding probability density can be derived in discrete form from the binned data according to the formula

$$p_j = \frac{m_j}{M\Delta}, \tag{J.2}$$

where m_j is the count of samples in bin j, M is the sum across all of the bin values (using $M = 5000$) and Δ is the bin-width in x (here about 2.5 metres).

To clarify where this equation comes from, refer back to Section 4.2 where it is stated that the probability P of finding a value of x in the small range $[x, x + dx]$ is given by $P = p(x) \, dx$, where $p(x)$ is the probability density. In the present case, we have a discrete set of data from which we are attempting to approximate an underlying continuous distribution.

On this basis, using equation J.2 to stand for $p(x)$, we have

$$P = p_j \Delta,$$
$$= \frac{m_j}{M},$$

which is, as might be expected, just the proportion of binned values that fall in the interval $[x, x+\Delta]$. This is consistent with our understanding of a probability involving discrete quantities.

We can now plot the computed probability densities for both x_n and y_n using equation (J.2), and compare the results to the theoretical normal density, giving Figure J.2.

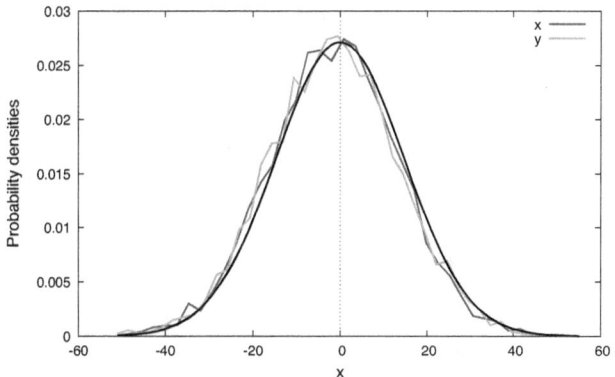

Figure J.2: Probability densities for x and y at $n = 600$

The smooth black curve stands for equation (J.1), using $s_n = 14.697$ m (which is near enough the same value for both x and y).

This is not quite the end of the matter, though, even after having gained confidence in the appropriate forms of $p(x)$ and $p(y)$. What we are after is the probability P_R that a mosquito will be found within a radius of R from its starting point after n steps. Or, conversely, we want the (hopefully small) probability that the mosquito will be *outside* that radius after n steps. The former quantity involves integrating the product $p(x)p(y)$ over the area Ω within the circle, so that

$$P_R = \int_\Omega p(x)p(y)\,dx\,dy,$$

where the circular area Ω includes all values of x and y for which a radial distance r from the origin is constrained by $r \leq R$, with $r = \sqrt{x^2 + y^2}$. Bolt in the normal forms for the probability densities from equation (J.1) to get:

$$P_R = \int_\Omega \frac{1}{s_n\sqrt{2\pi}} e^{-x^2/2s_n^2} \frac{1}{s_n\sqrt{2\pi}} e^{-y^2/2s_n^2}\,dx\,dy,$$

$$= \frac{1}{2\pi s_n^2} \int_\Omega e^{-(x^2+y^2)/2s_n^2}\,dx\,dy.$$

Since the area Ω is defined in polar coordinates by the equation $r \leq R$, and there appears an $x^2 + y^2 = r^2$ in the exponential, it makes sense to change to polar coordinates r and θ. Under this change in coordinates, the element of area $dx\,dy$ maps to $r\,dr\,d\theta$ [6], so we get

$$P_R = \frac{1}{2\pi s_n^2} \int_{\theta=0}^{2\pi} \int_{r=0}^{R} r e^{-r^2/2s_n^2}\,dr\,d\theta,$$

$$= \frac{1}{s_n^2} \int_{r=0}^{R} r e^{-r^2/2s_n^2}\,dr,$$

using the fact that the integrand is independent of θ and that $\int_{\theta=0}^{2\pi} d\theta = 2\pi$. The integral over r is straightforward enough to deal with, since

$$\frac{d}{dr}\left(e^{-r^2/2s_n^2}\right) = -\frac{r}{s_n^2}e^{-r^2/2s_n^2},$$

which means that

$$\begin{aligned} P_R &= \int_{r=0}^{R} \left(-\frac{d}{dr}e^{-r^2/2s_n^2}\right) dr, \\ &= \left[-e^{-r^2/2s_n^2}\right]_{r=0}^{R}, \\ &= 1 - e^{-R^2/2s_n^2}. \end{aligned} \quad \text{(J.3)}$$

Therefore, the complementary probability $\overline{P_R}$ that the mosquito will be found *outside* the circle after n steps must be

$$\overline{P_R} = 1 - P_R = e^{-R^2/2s_n^2}.$$

It can be seen that the functional dependence is on the ratio $\xi = R/s_n$, which simplifies calculations. That is, since $s_n = \sigma\sqrt{n}$ (see above) the key quantity involved in the probabilities is just

$$\xi = \frac{R}{\sigma\sqrt{n}}.$$

We are still not quite finished — the final step is to make sure that the simulated data is actually consistent with equation (J.3). This is achieved quite simply by counting every value of x-y pairs (at the 600-step mark) for which $\sqrt{x^2 + y^2} > R$, for a series of different R-values, and then computing the ratio of that count relative to the total number of samples (namely 5000). The result is shown as the set of crosses in Figure J.3, in comparison with the theoretical curve $e^{-\xi^2/2}$ (which is the green line).

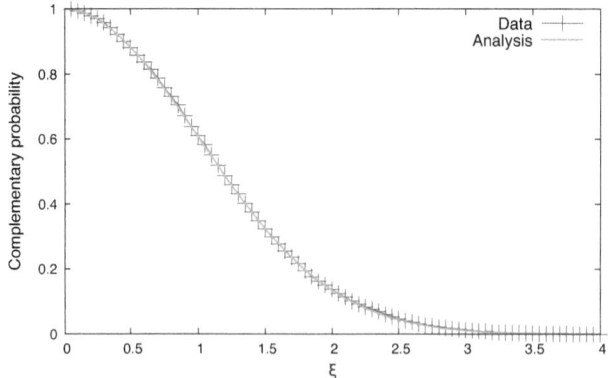

Figure J.3: Probability of $r > R$ at $n = 600$

This can be used to infer that there is only a very small probability ($P_R = 3.35 \times 10^{-4}$) of finding a mosquito at some distance $4 \times 14.696 = 58.784$ m from its starting point after 600 steps.

As mentioned earlier, this simulation is entirely notional so far as mosquito capabilities are concerned. And any prevailing breeze will alter the statistics considerably, since the motion will no longer be a pure random walk (an interesting problem in its own right, but I won't get into that).

Appendix K

Bending Beams

CHAPTER 12 provided a general equation for the vertical displacement y of a uniform beam under a continuous load $f(x)$ per unit length:

$$EI\frac{d^4y}{dx^4} = f(x),$$

assuming that the quantity EI is a constant. The convention used in this equation is that y is measured positive downwards and f also acts positive downwards. However, to make the illustrated deflection graphs conform to a more usual reader expectation, it is sensible to define both y and f as positive upwards instead. The above equation is unaltered.

This appendix goes through a couple of selected examples to show how this equation — or its constituent predecessors where relevant — may be applied in practice.

K.1 Cantilever Beam Bending Under Its Own Weight

One obvious example of a continuous load is when a cantilever beam is bending under its own weight, in which case

$$f(x) = -\rho A g,$$

where ρ is the density, A the cross-sectional area and g the downward (and so negative) acceleration due to gravity[1]. Therefore,

$$EI\frac{d^4y}{dx^4} = -\rho A g. \tag{K.1}$$

A quick cross-check on the dimensions involved here can provide confidence that this equation actually makes dimensional sense. Young's modulus E is known

---

[1] The mass of a small slice dx of the beam would be $\rho A dx$, but it is the force per unit length that is needed for the above equation.

to have units of Newtons per square metre, usually written as Nm^{-2}. Let M stand for mass, X for length and T for time [38]; then, dimensionally, a Newton is given as:

$$N \sim \frac{MX}{T^2},$$

here using the symbol $\sim$ to stand for 'has dimensions of'. So,

$$E \sim \frac{MX}{T^2}\frac{1}{X^2} \sim \frac{M}{XT^2}.$$

Similarly, $I = \iint z^2 dA$ has dimensions X^4, resulting in

$$EI \sim \frac{MX^3}{T^2}.$$

The left-hand side of equation (K.1) thus has dimensions

$$\text{left-hand-side} \sim \frac{MX^3}{T^2}\frac{X}{X^4} \sim \frac{M}{T^2},$$

since d^4y has dimensions X, while dx^4 has dimensions X^4.

Now look at the right-hand-side of equation (K.1). Its constituents have the following dimensions:

$$\rho \sim \frac{M}{X^3} \text{ (mass per unit volume)}; \quad A \sim X^2 \text{ (area)}; \quad g \sim \frac{X}{T^2} \text{ (acceleration)}.$$

Putting this lot together:

$$\text{right-hand-side} \sim \frac{M}{X^3}X^2\frac{X}{T^2} \sim \frac{M}{T^2}.$$

So equation (K.1) is at least dimensionally consistent.

To simplify the subsequent algebra, write

$$\frac{EI}{\rho A g} = \kappa,$$

giving

$$\kappa \frac{d^4y}{dx^4} = -1,$$

and integrate once to get:

$$\kappa \frac{d^3y}{dx^3} = \alpha - x,$$

for constant α. Borrowing from Section 25.4, the curvature and rate of change of curvature of the beam are both zero at the free end, where x equals L. This gives $\alpha = L$. Then integrate again and apply the free end condition, to get

$$\kappa \frac{d^2y}{dx^2} = -\frac{L^2}{2} + Lx - \frac{1}{2}x^2.$$

Integrate this once more:

$$\kappa \frac{dy}{dx} = \beta - \frac{L^2 x}{2} + \frac{1}{2}Lx^2 - \frac{1}{6}x^3.$$

For a cantilever beam, both y and dy/dx must be zero at $x = 0$, giving $\beta = 0$. Integrating for the last time and applying $y = 0$ at $x = 0$ then finally results in:

$$\kappa y = -\frac{x^2}{24}\left\{6L^2 - 4Lx + x^2\right\},$$

after a bit of rearranging.

But what does the actual shape of the bent beam look like? The measurable vertical displacement will depend on the length, density, Young's modulus and cross-sectional shape and area, but a typical profile would look like Figure K.1.

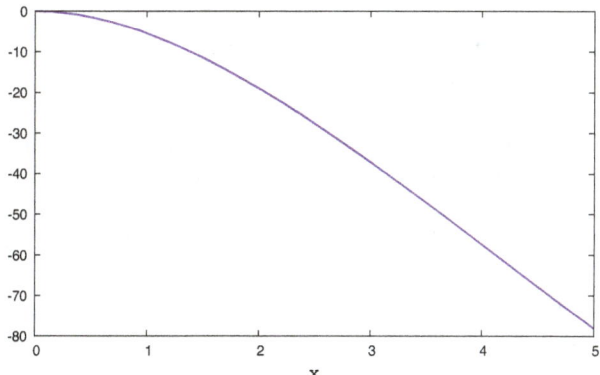

Figure K.1: Beam bent under uniform load

In this case, the vertical axis is κy and the beam length L is 5 metres. As might have been anticipated, the maximum deflection occurs at the free end, in which case y_{max} is given by:

$$\kappa y_{max} = -\frac{L^4}{8}.$$

K.2 Cantilever Beam With Single Point Load

A different approach is needed if the load is discrete — namely, concentrated at some point a from the left-hand-end. A cantilever beam is still assumed, but now of sufficiently light construction that its own mass can be neglected. Equation (12.13) is now unsuitable as a starting point, since (from the discussion in Section 12) the derivative of M is discontinuous, as are the third and higher derivatives of y. This can be understood, since the load itself is discontinuous, being zero everywhere along the beam except at $x = a$.

In this case, it helps to consider what is going on from near first principles, so refer to Figure K.2.

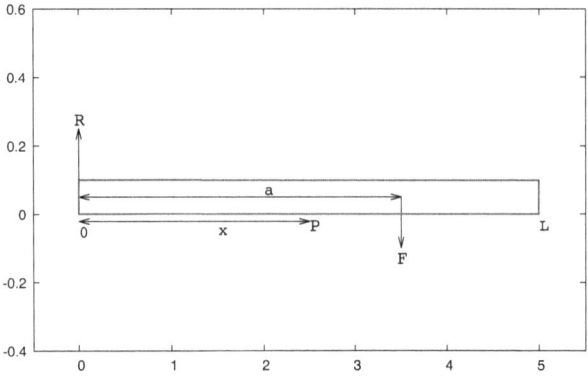

Figure K.2: Beam bent under a point load

Here, the beam of length L has a single downward-pointing load F imposed at a distance a from the left-hand end. The beam is assumed set into a wall on the left, meaning it cannot move and cannot bend at that point.

The upward-pointing reaction force R is needed to equalise the downward F, since the beam is in equilibrium. Its value can be obtained from a simple sum of forces, as discussed in Section 12, giving $R = F$. In more complex situations, where several loads are applied, it is often necessary to work out the various reaction forces by taking moments about the points of support [33], but we do not need that elaboration here.

The shear force is now obtained from

$$S = -\sum_i F_i,$$

and — as mentioned in Section 12 — can be expected to be discontinuous. So moving the point P along the beam from left to right will start out having $S = -R$ and then change to $S = 0$ after $x = a$, since the forces then cancel out. In equation form:

$$S = \begin{cases} -R & \text{for } 0 \leq x < a, \\ 0 & \text{for } a \leq x \leq L. \end{cases}$$

We can now work from

$$\frac{dM}{dx} = -S,$$

(see Section 12), although it is now necessary to split the solution into two parts.

- $0 \leq x < a$:

    Here, $\dfrac{dM}{dx} = R$, giving the solution by integration as $M = Rx + \alpha$.

K.2. CANTILEVER BEAM WITH SINGLE POINT LOAD

- $a \leq x \leq L$:

  For this part of the beam, $S = 0$, so $\dfrac{dM}{dx} = 0$, which implies that $M = \beta$.

And α and β are some constants to be determined.

Invoke equation (12.12), namely

$$M = EI\frac{d^2y}{dx^2},\qquad\text{(K.2)}$$

to relate M to the curvature in y, with y measured positive upwards. For a cantilever beam, the free end at L will have zero for both the second and third derivatives of y, which immediately implies that $\beta = 0$. Imposing continuity of M across the discontinuity at $x = a$ then gives $\alpha = -Ra$.

To summarise, the moment M at any point x along the beam will be:

$$M = \begin{cases} R(x-a) & \text{for } 0 \leq x < a, \\ 0 & \text{for } a \leq x \leq L. \end{cases}$$

The rest of the solution requires equation (K.2) to be solved for y, treating the regions either side of a separately, imposing $y = 0$ and $dy/dx = 0$ at $x = 0$, and making y and dy/dx continuous across $x = a$. The results are as follows:

$$\frac{EIy}{R} = \begin{cases} \frac{x^2}{6}(x-3a) & \text{for } 0 \leq x < a, \\ \frac{a^2}{6}(a-3x) & \text{for } a \leq x \leq L. \end{cases}$$

So a cubic curve for $x < a$ and linear thereafter. The profile for y looks like Figure K.3, here using $L = 5$ and $a = 3.5$. The y axis plots EIy/R in this case.

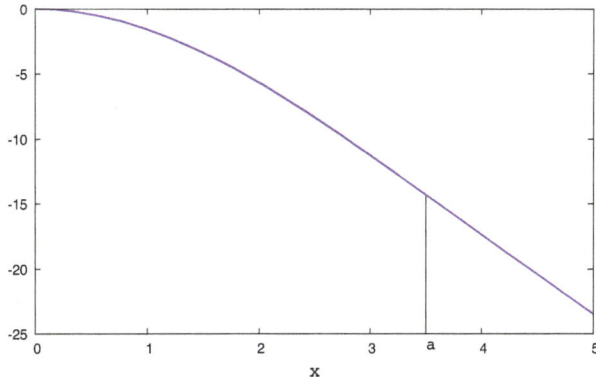

Figure K.3: Beam bent under a discrete load at a

As a final remark here, the inclusion of beam mass into the above discrete load formulation (*i.e.* a heavy beam plus one or more point loads) requires a re-evaluation of S to take account of the mass per unit length, but still incorporating the discontinuities [33].

Appendix L

Derivation of Shallow-Water Equations

SECTION 27 used the linearised shallow-water partial differential equations to obtain the speed of propagation of a tsunami wave over the ocean surface. These equations were quoted without much explanation or derivation, so this appendix attempts to fill the gap in understanding.

Start from the foundational Navier-Stokes equations [37], on the assumption that the only body-force is gravity acting downwards and that the fluid is incompressible (a good approximation for water) and frictionless (*i.e.* negligible viscosity, meaning that water molecules slide fairly freely past each other):

$$\frac{\partial u}{\partial x} + \frac{\partial v}{\partial y} + \frac{\partial w}{\partial z} = 0, \tag{L.1}$$

$$\frac{\partial u}{\partial t} + u\frac{\partial u}{\partial x} + v\frac{\partial u}{\partial y} + w\frac{\partial u}{\partial z} = -\frac{1}{\rho}\frac{\partial p}{\partial x}, \tag{L.2}$$

$$\frac{\partial v}{\partial t} + u\frac{\partial v}{\partial x} + v\frac{\partial v}{\partial y} + w\frac{\partial v}{\partial z} = -\frac{1}{\rho}\frac{\partial p}{\partial y}, \tag{L.3}$$

$$\frac{\partial w}{\partial t} + u\frac{\partial w}{\partial x} + v\frac{\partial w}{\partial y} + w\frac{\partial w}{\partial z} = -\frac{1}{\rho}\frac{\partial p}{\partial z} - g. \tag{L.4}$$

In addition, there are boundary conditions to be taken into account:

On $z = -h(x,y)$ (the sea floor): $u\frac{\partial}{\partial x}(z+h(x,y)) + v\frac{\partial}{\partial y}(z+h(x,y))$

$$+ w\frac{\partial}{\partial z}(z+h(x,y)) = 0, \tag{L.5}$$

On the surface at $z = \eta(x,y,t)$: $w = \frac{\partial \eta}{\partial t} + u\frac{\partial \eta}{\partial x} + v\frac{\partial \eta}{\partial y} + w\frac{\partial \eta}{\partial z}, \tag{L.6}$

Also on the surface at $z = \eta(x,y,t)$, the pressure $p = 0$.

The coordinates are such that x and y are in the horizontal plane (aligned with an undisturbed sea surface), while z is measured upwards from the mean

APPENDIX L. DERIVATION OF SHALLOW-WATER EQUATIONS

sea surface level. The velocity components u, v and w are in the x, y and z directions respectively.

The sea floor is modelled by the function $h(x, y)$, assumed time-independent (and actually negative given the choice of the z axis), while $\eta(x, y, t)$ models the shape of the sea surface (necessarily time-dependent).

Equation (L.5) reflects the fact that the sea floor is assumed to be impenetrable — the water velocity cannot have a component normal to $z = -h(x, y)$ (see Section 14.1 for a further discussion on surface normal vectors).

Equation (L.6) says that the vertical fluid velocity component at the sea surface must be equal to the kinematic rate of change of the wave shape $\eta(x, y, t)$. Also at the surface, the water pressure matches that of the atmosphere, which is assumed to be negligible in comparison with the water pressure. The effects of surface tension are neglected.

Now examine the global conservation of mass by integrating equation (L.1) over the water depth:

$$0 = \int_{z=-h}^{\eta} \left(\frac{\partial u}{\partial x} + \frac{\partial v}{\partial y} + \frac{\partial w}{\partial z} \right) dz,$$

$$= \int_{z=-h}^{\eta} \frac{\partial u}{\partial x} dz + \int_{z=-h}^{\eta} \frac{\partial v}{\partial y} dz + [w]_{z=\eta} - [w]_{z=-h},$$

$$= \int_{z=-h}^{\eta} \frac{\partial u}{\partial x} dz + \int_{z=-h}^{\eta} \frac{\partial v}{\partial y} dz + \left[\frac{\partial \eta}{\partial t} + u\frac{\partial \eta}{\partial x} + v\frac{\partial \eta}{\partial y} + w\frac{\partial \eta}{\partial z} \right]_{z=\eta}$$

$$- [w]_{z=-h}, \tag{L.7}$$

using equation (L.6) to obtain the penultimate term here.

Now expand out the boundary condition, equation (L.5), resulting in:

$$\left[u\frac{\partial h}{\partial x} + v\frac{\partial h}{\partial y} + w \right]_{z=-h} = 0,$$

using the fact that $h(x, y)$ is independent of z. Rearranging and substituting into equation (L.7) then gives:

$$0 = \int_{z=-h}^{\eta} \frac{\partial u}{\partial x} dz + \int_{z=-h}^{\eta} \frac{\partial v}{\partial y} dz + \left[\frac{\partial \eta}{\partial t} + u\frac{\partial \eta}{\partial x} + v\frac{\partial \eta}{\partial y} + w\frac{\partial \eta}{\partial z} \right]_{z=\eta}$$

$$+ \left[u\frac{\partial h}{\partial x} + v\frac{\partial h}{\partial y} \right]_{z=-h}. \tag{L.8}$$

To deal with the first two integrals in this equation, we will need the following representative result:

$$\frac{\partial}{\partial x} \int_{z=a(x)}^{b(x)} u(x, z) dz = \int_{z=a(x)}^{b(x)} \frac{\partial u(x, z)}{\partial x} dz + \frac{\partial b(x)}{\partial x} [u(x, z)]_{z=b(x)}$$

$$- \frac{\partial a(x)}{\partial x} [u(x, z)]_{z=a(x)},$$

which means that

$$\int_{z=-h}^{\eta} \frac{\partial u}{\partial x} dz = \frac{\partial}{\partial x} \int_{z=-h}^{\eta} u\, dz - \frac{\partial \eta}{\partial x} [u]_{z=\eta} - \frac{\partial h}{\partial x} [u]_{z=-h}.$$

Substituting this and its companion for $\partial v/\partial y$ into equation (L.8) results in:

$$\begin{aligned} 0 = & \frac{\partial}{\partial x} \int_{z=-h}^{\eta} u\, dz - \frac{\partial \eta}{\partial x} [u]_{z=\eta} - \frac{\partial h}{\partial x} [u]_{z=-h} \\ & + \frac{\partial}{\partial y} \int_{z=-h}^{\eta} v\, dz - \frac{\partial \eta}{\partial y} [v]_{z=\eta} - \frac{\partial h}{\partial y} [v]_{z=-h} \\ & + \left[\frac{\partial \eta}{\partial t} + u \frac{\partial \eta}{\partial x} + v \frac{\partial \eta}{\partial y} + w \frac{\partial \eta}{\partial z} \right]_{z=\eta} \\ & + \left[u \frac{\partial h}{\partial x} + v \frac{\partial h}{\partial y} \right]_{z=-h}. \end{aligned}$$

Since $\partial h/\partial x$ and $\partial h/\partial y$ are independent of z, the last two terms here cancel with the third and sixth terms on the right-hand side, thus giving:

$$\begin{aligned} 0 = & \frac{\partial}{\partial x} \int_{z=-h}^{\eta} u\, dz - \frac{\partial \eta}{\partial x} [u]_{z=\eta} + \frac{\partial}{\partial y} \int_{z=-h}^{\eta} v\, dz - \frac{\partial \eta}{\partial y} [v]_{z=\eta} \\ & + \left[\frac{\partial \eta}{\partial t} + u \frac{\partial \eta}{\partial x} + v \frac{\partial \eta}{\partial y} + w \frac{\partial \eta}{\partial z} \right]_{z=\eta}. \end{aligned}$$

In a similar manner, the two central terms in the last square-bracketed quantity cancel with the second and fourth terms on the right-hand side. Also, the last term in the square brackets is zero since η is independent of z, so that:

$$0 = \frac{\partial}{\partial x} \int_{z=-h}^{\eta} u\, dz + \frac{\partial}{\partial y} \int_{z=-h}^{\eta} v\, dz + \left[\frac{\partial \eta}{\partial t} \right]_{z=\eta}.$$

And, again, as η depends only on x, y and t, the $z = \eta$ condition on the last term is irrelevant and the brackets can be removed:

$$0 = \frac{\partial}{\partial x} \int_{z=-h}^{\eta} u\, dz + \frac{\partial}{\partial y} \int_{z=-h}^{\eta} v\, dz + \frac{\partial \eta}{\partial t}. \tag{L.9}$$

We are nearly (but not quite) at the end of the derivation. Return now to equation (L.4) and assume that the vertical acceleration of the water is sufficiently small that it can be ignored, in which case the right-hand-side of that equation simplifies to:

$$\frac{\partial p}{\partial z} = -\rho g.$$

This can be integrated directly, giving

$$p(x, y, z, t) = p_0(x, y, t) - \rho g z.$$

APPENDIX L. DERIVATION OF SHALLOW-WATER EQUATIONS

Application of the free-surface condition on p gives $p_0 = \rho g\, \eta(x,y,t)$, so that

$$p(x,y,z,t) = \rho g\left(\eta(x,y,t) - z\right). \tag{L.10}$$

With this in hand, the right-hand-sides in equations (L.2) and (L.3) can be evaluated. In addition, on the assumption that the vertical z-motion of the fluid can be ignored, the following two nonlinear equations for u and v result:

$$\frac{\partial u}{\partial t} + u\frac{\partial u}{\partial x} + v\frac{\partial u}{\partial y} + g\frac{\partial \eta}{\partial x} = 0, \tag{L.11}$$

$$\frac{\partial v}{\partial t} + u\frac{\partial v}{\partial x} + v\frac{\partial v}{\partial y} + g\frac{\partial \eta}{\partial y} = 0. \tag{L.12}$$

The final step is to assume that both u and v are independent of z, which then simplifies the integrals in equation (L.9). That is,

$$\frac{\partial}{\partial x}\int_{z=-h}^{\eta} u(x,y,t)\,dz = \frac{\partial}{\partial x}\left\{u(x,y,t)\,[z]_{z=-h}^{\eta}\right\},$$

$$= \frac{\partial}{\partial x}\left\{u(x,y,t)\Big(\eta(x,y,t) + h(x,y)\Big)\right\},$$

and similarly for the y-integral. Therefore,

$$\frac{\partial \eta}{\partial t} + \frac{\partial}{\partial x}\left[u\left(\eta + h\right)\right] + \frac{\partial}{\partial y}\left[u\left(\eta + h\right)\right] = 0. \tag{L.13}$$

Equations (L.11), (L.12) and (L.13) form the shallow-water set.

Appendix M

Parallel Displacement

THE left-hand-side of equation (29.1) at the start of Chapter 29, namely

$$R_{\mu\nu} - \frac{1}{2}g_{\mu\nu}R + \Lambda g_{\mu\nu},$$

was presented to the reader as describing 'geometry', but without any accompanying justification for the term. And I had intended to leave it at that, rather than embark on a lengthy series of explanatory steps. However, second thoughts suggested that it might be helpful to add at least a bit of background, even if sketchy and incomplete, and the following paragraphs are my attempt at this. To fill in the gaps (and there are quite a lot of gaps), I do recommend reading more dedicated references.

Firstly, a couple of introductory definitions and conventions: a tensor is a quantity that transforms from one coordinate system to another in a well-defined manner, and there are two types:

$$\text{Contravariant: } \bar{A}^p = \frac{\partial \bar{x}^p}{\partial x^q} A^q,$$

$$\text{Covariant: } \bar{B}_p = \frac{\partial x^q}{\partial \bar{x}^p} B_q.$$

In both cases, the vectors A^p and B_p are transformed from coordinate system x to another coordinate system $\bar{x}$. The top and bottom indices are coordinate markers (and not powers in the superscripted case), so (for example) A^q could stand for the three-component set (A^1, A^2, A^3). The placement of the index, superscript or subscript, indicates which tensor type we are dealing with (the distinction is not actually important for what follows below, but I might as well fill this particular gap).

In a similar manner, x^p can (for instance) represent the set x^1, x^2, x^3, standing in for polar coordinates r, θ, ϕ. Coordinates are always superscripted.

By Einstein's summation convention, twice-repeated indices are summed over their respective range, thus obviating the need for the $\sum$ symbology and resulting in more compact equations.

Higher-order tensors transform between coordinate systems in a similar manner.

Tensors have been introduced here because they have an important property: a tensor equation that is valid in one coordinate system is valid in all coordinate systems; the value of this will become apparent later.

Most of us who have had some background in physics, or enjoy reading books on popular science, will be aware that our experience of gravity is actually a consequence of curved space-time. And it should be self-evident that since we inhabit this space-time, we can't get outside it to 'see' such curvature directly (whatever 'seeing' might mean in this context). This insight highlights the difference between extrinsic and intrinsic curvature. The former term refers to assessing the curvature of some surface, say, when observed from the outside — as when measuring the characteristics of a sphere or other shape placed on a workbench. We can't do that for space-time: our position is similar to that of an ant (say) confined to the two-dimensional surface of a sphere, and constrained to making measurements only on that surface. Such an ant would therefore be dealing with intrinsic curvature.

The methods of differential geometry were developed with this requirement in mind, and the concept of parallel displacement is key to characterising curvature. To get a feel for what's going on, suppose we confine ourselves to two spatial dimensions and start with a flat space — a table-top or whatever — and place a small arrow flat on the surface. Now move the arrow around on the surface in a series of small steps, at each step keeping the arrow parallel to itself. There should be no great surprise to find that going around a closed circuit ends up with the arrow parallel to its starting orientation.

Doing the same thing on a *curved* surface will in general give different results, such that the starting and ending arrows are no longer in alignment. To proceed mathematically, though, some clarifications are needed as to what we actually mean by 'parallel' in the general case. Visualise what is going on by means of a thought experiment: start with the arrow tangential to this curved surface, and parallel-move it a bit to a new location. It will then cease to be tangential, but adjusting its angle down a bit (and only downwards toward the surface) will retrieve its tangential property. This somewhat hand-waving explanation gives an insight into how the concept of parallel transport can be generalised.

In mathematical terms, the equation for parallel displacement of a vector ξ embedded in a metric space is written in the form [90]:

$$d\xi^\mu = - \left\{ {\mu \atop \alpha\ \beta} \right\} dx^\alpha \xi^\beta, \tag{M.1}$$

where ξ^β is the vector in question, $d\xi^\mu$ is its small displacement, dx^α stands for a small change in the relevant coordinate, and

$$\left\{ {\mu \atop \alpha\ \beta} \right\} = g^{\mu k} \left[\alpha\,\beta,\ k \right],$$

is the Christoffel[1] symbol of the second kind (symmetric in an interchange of

---
[1] Elwin Christoffel, 1829 to 1900.

the lower two indices). The associated quantity here,

$$[\alpha\beta, k] = \frac{1}{2}\left(\frac{\partial g_{\alpha k}}{\partial x^\beta} + \frac{\partial g_{\beta k}}{\partial x^\alpha} - \frac{\partial g_{\alpha\beta}}{\partial x^k}\right\},$$

is known as the Christoffel symbol of the first kind (also symmetric in α and β).

Or, writing out equation (M.1) in full:

$$d\xi^\mu = -\frac{1}{2}g^{\mu k}\left(\frac{\partial g_{\alpha k}}{\partial x^\beta} + \frac{\partial g_{\beta k}}{\partial x^\alpha} - \frac{\partial g_{\alpha\beta}}{\partial x^k}\right)dx^\alpha \xi^\beta, \qquad (M.2)$$

where the g_{pq} entries are the components of the metric tensor defining the local geometry, so that

$$ds^2 = g_{pq}dx^p dx^q,$$

(and the superscript ^{2} does here mean ds-squared).

The contravariant $g^{\mu k}$ quantity in equation (M.2) is defined in term of the metric tensor via the following relation:

$$g^{\mu k}g_{\nu k} = \delta^\mu_\nu,$$

where δ^μ_ν is the Kronecker[2] delta: that is, unity if $\mu = \nu$ and zero otherwise. And for future reference, $g_{\mu\nu}$ and $g^{\mu\nu}$ are used more generally to raise and lower indices in tensor equations.

Going into the detail of where equation (M.1) comes from would take up too much space in this book. However, a couple of criteria that are involved can be mentioned here to provide some understanding. Firstly, the transplantation law should be coordinate-invariant (meaning that the transported vector remains a vector). And secondly, the vector that is being transported needs to keep its physical length unchanged[3]. This, in turn, requires that the quantity

$$g_{\mu\nu}\xi^\mu\xi^\nu$$

must be constant during the transport operation. This entity is the square of the *physical* length of the vector, and is different from the summed-squares of the coordinate components. The distinction between the two lengths is perhaps best appreciated by means of a simple example from polar coordinates, in which

$$ds^2 = dr^2 + r^2 d\theta^2.$$

Thus the physical length in the angle direction is given by $rd\theta$ and not $d\theta$ by itself.

It is evident that if the g_{pq} components are constant, then their derivatives will be zero and $d\xi^\mu = 0$, so vector ξ^β is displaced unchanged; this is as would be expected in flat space.

---

[2] Leopold Kronecker, 1823 to 1891.
[3] More generally, the scalar product of two vectors remains invariant when parallel-transported.

Things get a bit more interesting on the surface of a sphere of radius R, for which

$$ds^2 = R^2 \left(d\theta^2 + \sin^2 \theta \, d\phi^2 \right), \tag{M.3}$$

the angle θ being the co-latitude, measured downwards from the north polar axis, while ϕ is measured in the equatorial plane. Define $dx^1 = \theta$ and $dx^2 = \phi$, with some arbitrary vector $\xi^\mu = (p, q)$ as the arrow tangential to the surface. Given the above metric, the only non-zero Christoffel symbols are as follows [86]:

$$\begin{Bmatrix} 1 \\ 2\ 2 \end{Bmatrix} = -\sin\theta\cos\theta; \quad \begin{Bmatrix} 2 \\ 1\ 2 \end{Bmatrix} = \begin{Bmatrix} 2 \\ 2\ 1 \end{Bmatrix} = \cot\theta. \tag{M.4}$$

Therefore, equation (M.1) becomes:

$$dp = \sin\theta\cos\theta \, q \, d\phi,$$
$$dq = -\cot\theta \, (q\, d\theta + p\, d\phi)$$

These equations define the twisting of the vector ξ^μ as it is moved over some surface path.

Suppose, by way of example, that the path chosen is a closed circle of constant latitude, a choice that renders the calculation straightforward. That is, set $\theta = \theta_0$ so that $d\theta = 0$, while varying ϕ; the above equations for dp and dq become:

$$\frac{dp}{d\phi} = \sin\theta_0 \cos\theta_0 \, q, \tag{M.5}$$

$$\frac{dq}{d\phi} = -\cot\theta_0 \, p. \tag{M.6}$$

These can be combined into a single second-order differential equation in p, namely:

$$\frac{d^2 p}{d\phi^2} + \omega^2 p = 0,$$

which describes periodic variability in p with an associated frequency

$$\omega = \cos\theta_0,$$

and with q having the same functional form.

For an example of what the results look like, see Figure M.1, showing the upper half of a sphere of radius 100 m.

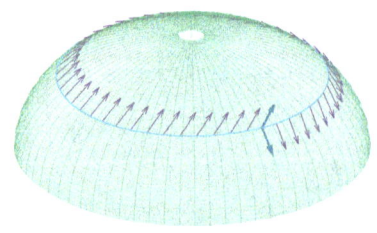

Figure M.1: Illustration of parallel transport

The start and end points of the vector migration are coloured dark green, with the starting orientation pointing downwards toward the equator. The closed path at $\theta_0 = 50°$ is traversed in an anti-clockwise direction, and it can be seen that the final vector ends up pointing north-east rather than south. For visual clarity, the initial vector was given a length of 20 m which (as expected) remained constant throughout.

Only for a path around the equator, for which $\theta_0 = \pi/2$, does vector ξ^β migrate unchanged.

Deriving the arrows in Figure M.1 required mapping the angular vector θ-ϕ coordinates for the start and end of each arrow, namely (θ_0, ϕ) and $(\theta_0 + p, \phi + q)$ respectively, into Cartesian form consistent with the half-sphere shown.

For a more general path over the surface, described in terms of $\theta(s)$, $\phi(s)$ for path length parameter s, the equations to be solved would be:

$$\frac{dp}{ds} = \sin\theta \cos\theta \frac{d\phi}{ds} q, \tag{M.7}$$

$$\frac{dq}{ds} = -\cot\theta \left(q \frac{d\theta}{ds} + p \frac{d\phi}{ds} \right). \tag{M.8}$$

In this general case, verifying that the physical vector length is propagated unchanged is a matter of determining that the derivative of

$$L^2 = p^2 + q^2 \sin^2\theta$$

with respect to s is zero. Carrying out the derivatives and bolting in equations (M.7) and (M.8) does then result in $dL/ds = 0$.

It can be appreciated that the results of parallel-transporting a vector to a new location will depend on the particular path taken; we'll come back to this point later on.

Thus far, we have only looked at parallel displacement, which is a rather specific form of vector-shifting. Now consider a more general small change in coordinates, by some dx^j, say. From a Taylor series expansion, the value of the vector at that shifted position then will be given by:

$$\xi^i \left(x^j + dx^j \right) = \xi^i \left(x^j \right) + \frac{\partial \xi^i}{\partial x^k} dx^k + O\left(dx^k\right)^2.$$

Compare this to a parallel displacement of the same form, namely:

$$\xi^{i*} \left(x^j + dx^j \right) = \xi^i \left(x^j \right) - \left\{ {i \atop k\ l} \right\} \xi^l dx^k + O\left(dx^k\right)^2,$$

and take the difference of the two shifted vectors:

$$\xi^i \left(x^j + dx^j \right) - \xi^{i*} \left(x^j + dx^j \right) = \left[\frac{\partial \xi^i}{\partial x^k} + \left\{ {i \atop k\ l} \right\} \xi^l \right] dx^k + O\left(dx^k\right)^2.$$

The quantity in square brackets on the right-hand-side has the form of a first derivative, and is termed the *covariant derivative* with respect to x^k of the vector ξ^i; it is here written as

$$\xi^i{}_{||k} = \frac{\partial \xi^i}{\partial x^k} + \left\{ {i \atop k\ l} \right\} \xi^l,$$

the double-bar symbology following that of [90].

This is the generalisation to tensors of the more familiar partial derivative, with the distinction that the covariant derivative is a tensor whereas the ordinary derivative is not. So to retain the tensorial nature of equations, the covariant derivative is important.

Now bring in higher derivatives, such as $\xi^i{}_{||k||m}$, meaning the covariant derivative of ξ^i with respect to x^k, followed by the covariant derivative with respect to x^m. A key difference from ordinary derivatives is that the two operations do not commute, so that

$$\xi^i{}_{||k||m} \neq \xi^i{}_{||m||k},$$

even though for ordinary derivatives the order does not matter.

This non-commutativity is used to define the Riemann[4] tensor $R^\alpha{}_{\eta\beta\gamma}$, using the equation:

$$\xi^\alpha{}_{||\beta||\gamma} - \xi^\alpha{}_{||\gamma||\beta} = R^\alpha{}_{\eta\beta\gamma}\xi^\eta.$$

It was mentioned above that parallel-displacing a vector to a new location will give a path-dependent result. If (for example) two different routes to the same final location are taken, it can be shown that the difference in the values obtained is proportional to the Riemann tensor.

This tensor $R^\alpha{}_{\eta\beta\gamma}$ is found to contain much important information regarding the curvature of a space, and is thus a natural choice to form a basis for the geometric requirements of general relativity. Einstein adopted the Ricci[5] tensor, formed from the Riemann tensor via the operation

$$R_{\mu\nu} = R^\alpha{}_{\mu\alpha\nu},$$

(with implied summation over the index α) and it is this tensor that appears in equation (29.1) (and at the start of this appendix). If this tensor is written out in terms of by-now-familiar quantities, we end up with

$$R_{\mu\nu} = \left\{\begin{array}{c}\alpha\\\mu\;\alpha\end{array}\right\}_{|\nu} - \left\{\begin{array}{c}\alpha\\\mu\;\nu\end{array}\right\}_{|\alpha} + \left\{\begin{array}{c}\alpha\\\tau\;\nu\end{array}\right\}\left\{\begin{array}{c}\tau\\\mu\;\alpha\end{array}\right\} - \left\{\begin{array}{c}\alpha\\\tau\;\alpha\end{array}\right\}\left\{\begin{array}{c}\tau\\\mu\;\nu\end{array}\right\}, \quad (M.9)$$

which is symmetric, in that $R_{\mu\nu} = R_{\nu\mu}$.

The subscripts $_{|\nu}$ and $_{|\alpha}$ used here stand for the usual partial derivatives with respect to x^ν and x^α respectively, so $R_{\mu\nu}$ involves both first and second derivatives of the metric tensor $g_{\mu\nu}$. The presence of *second* derivatives is consistent with the familiar mathematical treatment of acceleration in physics.

Fairly obviously, if all of the $g_{\mu\nu}$ components are constant, then $R_{\mu\nu} = 0$. But how about the metric $ds^2 = dr^2 + r^2 d\theta^2$? This actually describes flat space, but expressed in polar coordinate form (defined via $x = r\cos\theta$, $y = r\sin\theta$). The only non-zero Christoffel symbols are as follows:

$$\left\{\begin{array}{c}1\\2\;2\end{array}\right\} = -r; \quad \left\{\begin{array}{c}2\\1\;2\end{array}\right\} = \left\{\begin{array}{c}2\\2\;1\end{array}\right\} = \frac{1}{r},$$

---

[4] Georg Riemann, 1826 to 1866.
[5] After Gregorio Ricci-Curbastro, 1853 to 1925.

so it can be inferred that ξ^μ will in general rotate when parallel-transported. On the other hand, all components of $R_{\mu\nu}$ are found to be zero for this metric, so a zero Ricci tensor is indicative of flat space, regardless of the coordinate system used (the same property pertains to its parent, the Riemann tensor). As stated above, a tensor equation valid in one coordinate system is valid in all.

The final so-far-undefined quantity in equation (29.1) is the scalar R, which is defined as follows:

$$R = g^{\mu\nu} R_{\mu\nu}.$$

This is known as the Riemann curvature.

In brief, the above discussion has highlighted the relevance of the metric tensor $g_{\mu\nu}$ and the Christoffel symbols in the formulation of an equation for parallel transport. In turn, the concept of parallel transport provides information regarding the curvature of some surface — or, in more general terms, the curvature of some metric space — and further development of this idea leads to $R_{\mu\nu}$.

As a final remark here, if the thought of working out all of those Christoffel symbols seems daunting (certainly an error-prone process in my case), there is a 'back door' method based on deriving the geodesics. Recall from Chapter 23 that the curves of shortest path over a surface are given by minimising the integral

$$\int_a^b ds = \int_a^b \sqrt{g_{\mu\nu} dx^\mu dx^\nu} = \int_a^b \sqrt{g_{\mu\nu} \frac{dx^\mu}{ds} \frac{dx^\nu}{ds}} \, ds,$$

in terms of path length s. Use of the Euler-Lagrange equations then results[6] in the second-order differential equation,

$$\frac{d^2 x^r}{ds^2} + \begin{Bmatrix} r \\ \alpha \ \beta \end{Bmatrix} \frac{dx^\alpha}{ds} \frac{dx^\beta}{ds} = 0. \tag{M.10}$$

For any particular metric, deriving the Euler-Lagrange equations will result in a set of second-order equations. These can then be compared to equation (M.10), and the Christoffel symbols extracted directly. Try it out yourself for the metric $ds^2 = R^2 \left(d\theta^2 + \sin^2\theta \, d\phi^2 \right)$.

---

[6] Fairly obviously, I've left out the intermediate steps.

Appendix N

Greek Letters in Maths

THE Greek letters used in mathematics are listed in the table below, along with a phonetic equivalent to give some idea of pronunciation. The capital versions that are the same form as in the Roman alphabet have been omitted.

Lower case	Upper case	Phonetic form
α	–	alpha
β	–	beta
γ	Γ	gamma
δ	Δ	delta
ϵ	-	epsilon
ζ	-	zeta
η	-	eta
θ	Θ	theta
ι	-	iota
κ	-	kappa
λ	Λ	lambda
μ	-	mu
ν	-	nu
ξ	Ξ	xi
π	Π	pi
ρ	-	rho
σ	Σ	sigma
τ	-	tau
υ	Υ	upsilon
ϕ	Φ	phi
χ	-	chi
ψ	Ψ	psi
ω	Ω	omega

www.ingramcontent.com/pod-product-compliance
Lightning Source LLC
Chambersburg PA
CBHW052141220526
45471CB00004B/1465